Compressible Flow Propulsion and Digital Approaches in Fluid Mechanics

Mathematical and Mechanical Engineering Set

coordinated by
Abdelkhalak El Hami

Volume 4

Compressible Flow Propulsion and Digital Approaches in Fluid Mechanics

Michel Ledoux
Abdelkhalak El Hami

ISTE

WILEY

First published 2017 in Great Britain and the United States by ISTE Ltd and John Wiley & Sons, Inc.

ISTE Ltd
27-37 St George's Road
London SW19 4EU
UK

www.iste.co.uk

John Wiley & Sons, Inc.
111 River Street
Hoboken, NJ 07030
USA

www.wiley.com

Library of Congress Control Number: 2016959823

British Library Cataloguing-in-Publication Data
A CIP record for this book is available from the British Library
ISBN 978-1-78630-112-3

Contents

Preface

Mathematical physics was brought into existence by the development of mechanics. It originated in the study of the planetary motions and of the falling of heavy bodies, which had led Newton to formulate the fundamental laws of mechanics, as early as 1687. Even though the mechanics of continuous media, first as solid mechanics, and later as fluid mechanics, is a more recent development, its roots can be found in Isaac Newton's *Philosophiaenaturalis principia mathematica (Mathematical Principles of Natural Philosophy)*,several pages of which are dedicated to the falling streams of liquid.

Applications of fluid mechanics to irrigation problems date back to antiquity, but the subject gained a key status during the industrial revolution. Energetics was vital to the development of knowledge-demanding, specialized industrial areas such as fluid supply, heat engineering, secondary energy production or propulsion. Either as carrier of sensible heat or as core of energy production processes, fluid is ubiquitous in all the high-technology industries of the century: aeronautics, aerospace, automotive, industrial combustion, thermal or hydroelectric power plants, processing industries, national defense, thermal and acoustic environment, etc.

Depending on the target audience, there are various approaches to fluid mechanics. Covering this diversity is what we are striving for in this work.

Regardless of the degree of difficulty of the approached subject, it is important for the reader to reflect on it while being fully aware of the laws to be written in one form or another. Various approaches to fluid mechanics are illustrated by examples in this work.

First of all, the student will have the opportunity to handle simplified tools, allowing him/her a convenient first approach of the subject. On the other hand, the practitioner will be provided with elementary dimensioning means.

Other problems may justify or require a more complex approach, involving more significant theoretical knowledge, in particular of calculus. This is once again a point on which students and practitioners who already master these subjects can converge.

A third approach, which is essential for today's physics, especially when dealing with problems that are too complex to be accurately solved by simple calculations, resorts to numerical methods. This work illustrates these remarks.

Problem resolution relies in each chapter on reviews of fundamental notions. These reviews are not exhaustive, and the reader may find it useful to go back to textbooks for knowledge consolidation. Nevertheless, certain proofs referring to important points are resumed. As already mentioned, what matters is that the reader has a good grasp of what he/she writes.

Given that we target wide audiences, the deduction or review of general equations can be found in the appendices, to avoid the book becoming too cumbersome.

The attempt to effectively address audiences with widely varied levels of knowledge, expertise or experience in the field may seem an impossible task.

Drawing on their experience of teaching all these categories of audiences, the authors felt motivated and encouraged to engage in this daring enterprise.

Exploring subjects such as *propulsion, compressible flows, and the numerical approach to Fluid Mechanics,* this book primarily focuses on more complex problems, particularly on examples of industrial applications.

In Chapter 1 of this second volume, the subject of viscous fluids is resumed from an analytical perspective: the structure of flows at boundaries is taken into account; the boundary layer theory is extensively used; significant attention is given to integral methods; an introduction to this field in relation to non-Newtonian fluids is provided; viscous flows through piping systems are examined; some particular types of unsteady boundary layers are also addressed.

In high velocity fluid flows, the combination of thermodynamics and mechanics leads to behaviors that are highly characteristic of compressible fluids. Two chapters of this book focus on this part of fluid mechanics, a very important one especially in aeronautics and aerospace. The physics of nozzles, generators of jets producing the thrust of jet and rocket engines, is approached within the framework of one-dimensional flows.

Chapter 2 is dedicated to fully reversible, compressible one-dimensional flows.

Chapter 3 focuses on the emergence of local irreversibilities in such flows. The development of the theory of straight shock waves uses the example and the positioning of such shock waves in a nozzle.

The schematic and simplified form of flow is applicable to many practical cases and it validates the models described in the above mentioned chapters. In many cases, the high complexity of the flow renders oversimplification inappropriate. This is especially valid for turbulent or vortex flows in industrial applications. At present, the engineer or researcher has privileged access to computation tools covering a wide range of capacities.

Two chapters are dedicated to the numerical approach, which is required for problems involving structures that are too complex to be dealt with by elementary modeling.

Chapter 4 addresses modeling and simulation in fluid mechanics. Numerical computations are conducted in 2D, while flow variations in the vertical plane are neglected and blade tip losses are not taken into account.

The distribution of inlet wind velocity is considered uniform and the blades are considered straight.

Chapter 5 is dedicated to an industrial application. It refers to the performance evaluation of the straight-bladed Darrieus wind turbine by numerical computations and the validation of the numerical scheme by comparing the results obtained with the experimental results.

This work is aimed at students enrolled in engineering schools and technical colleges or in university Bachelor's or Master's programs. It is also meant to be useful to the professionals whose activity requires knowledge or mastery of tools related to fluid mechanics.

Michel LEDOUX
Abdelkhalak EL HAMI
November 2016

1

The Flow of Viscous Fluids. Flow in the Vicinity of a Wall: Boundary Layers and Films

1.1. Introduction

An important class of problems concerns the interaction between a viscous flow and a boundary: an open solid boundary (plate) or a closed one (tube), a fluid at rest, another flow or another fluid.

Given their common characteristics, these flows can be grouped in the category of boundary layers:

– *Outer boundary layers*: interaction between a flow of indefinite extension and a flat, curved, cylindrical or other shaped wall

– *Inner boundary layers*: interaction between a flow of finite extension and a partially or fully closed wall: pipe flow, open channel flow

– *"Mixing" boundary layers*: jets, etc.

These flows can be laminar or turbulent. In this respect, viscosity plays a key role.

1.2. Characteristics and classification of boundary layers

In essence, boundary layers are "thin" flows. Therefore, in any boundary layer, a distinction between "lateral" and "longitudinal" scales is established. In particular, the gradients of parameters (velocity, temperature, concentration) are much stronger along the lateral dimensions than along the longitudinal ones.

In the case of outer boundary layers, a two-step calculation method is applicable, which has already been mentioned in Chapter 3 of [LED 17]:

a) When a solid body is placed in a uniform flow, *the so-called "potential" flow is calculated* (see the kinematics review) based on a perfect inviscid fluid theory.

b) Viscosity in the "thin" boundary layer is then taken into account.

The "inner" boundary layers are present in areas where the regime sets in. Therefore, the scales to be compared are the radius of a pipe relative to the length L of this inflow area.

The jet boundary layers relate the lateral dimension of a jet, expressed by a lateral profile of velocity, and the longitudinal dimension of the jet.

The analytical approaches in this manual are focused on the outer or inner boundary layer. The numerical approach is better suited for jet problems.

A prototype of outer boundary layers consists of the flow developed by a flat plate in the uniform flow of velocity U_e [or locally uniform flow if $U_e = U_e(x)$ has an axial variation]. In this case, there is a non-disturbed flow (or "potential" flow; this problem has been mentioned in Chapter 3 of [LED 17]) and a border layer flow, where the flow connects to the wall area. This connection area with thickness δ is "thin" compared with the longitudinal dimension L of the body being considered.

There are longitudinal scales (along Ox in Cartesian coordinates), which are always larger than the lateral scales (along Oy in Cartesian coordinates).

Or expressed in orders of magnitude:

$$\delta << L \qquad [1.1]$$

There are two consequences for the *orders of magnitude* of the terms in the equations of fluid mechanics:

The boundary layer is "thin"; therefore, the flow is "nearly parallel to the flow". In Cartesian coordinates and for a plane flow, a first relation can be established between the longitudinal component u and lateral component v of velocity: $u << v$.

NOTE.– As will be seen, to ensure continuity, this cannot be rigorously valid. This will particularly be the case when employing integral methods. It will be used in certain cases as boundary "property" to obtain approximate results: see, for example, Stokes' first problem presented further on.

There is a similar relation, in terms of order of magnitude, between longitudinal and lateral derivatives of the same function f:

$$\frac{\partial f}{\partial x} << \frac{\partial f}{\partial y} \qquad [1.2]$$

A simplification of the equations of fluid mechanics is thus possible, though to a limited extent.

Indeed, in orders of magnitude:

$$\frac{\partial u}{\partial x} \approx \frac{u}{L}; \frac{\partial v}{\partial y} \approx \frac{v}{\delta}; v << u; \delta << L \qquad [1.3]$$

But: $\frac{u}{L} \approx \frac{v}{\delta}; \frac{\partial u}{\partial x} \approx \frac{\partial v}{\partial y}$ [1.4]

$$u\frac{\partial u}{\partial x} \approx u\frac{u}{L}; v\frac{\partial u}{\partial y} \approx u\frac{v}{\delta}; \frac{\partial^2 u}{\partial y^2} \approx \frac{u}{\delta^2} \qquad [1.5]$$

These three terms have the same order of magnitude.

For a plane incompressible steady flow above a flat plate, we have:

$$\frac{\partial u}{\partial x} + \frac{\partial v}{\partial y} = 0 \qquad [1.6]$$

$$u\frac{\partial u}{\partial x} + v\frac{\partial u}{\partial y} = \nu\frac{\partial^2 u}{\partial y^2} \qquad [1.7]$$

It is worth noting that not all the terms containing the derivative with respect to x disappear, as these equations contain products of small terms $\left(\frac{\partial u}{\partial x}, v\right)$ and large terms $\left(u, \frac{\partial u}{\partial y}, \frac{\partial^2 u}{\partial y^2}\right)$.

A physicist's approach to these order of magnitude-related notions is recommended. In the outer boundary layers, when the distance from the wall increases, velocity reaches quite rapidly a value close to that of the potential flow, U_e. Rigorously speaking, it is an asymptotic variation, meaning that u tends to U_e, when x tends to infinity. In wind tunnel, there would be infinities of several centimeters!

The thickness of the boundary layer, although denoted by δ, is not infinitely small. In wind tunnel applications, for a plate with length of several tens of centimeters, δ is also measured in centimeter. Given a ship with length $L = 250\, m$, δ is measured in meter. In a geographical area, the atmospheric boundary layer can measure 1–2 km!

Moreover, an infinitely small δ compared to L often comes down to ratios $\frac{\delta}{L}$ of the order of 10^{-2}.

1.2.1. *Boundary layers – various approaches*

For laminar flows, the boundary layer equations can be solved. Several examples, such as the Blasius theory and the Stokes theory, will be given below.

There are various approaches to turbulent flows: semi-empirical theories, the mixing length theory, the Boussinesq viscosity models and numerical approaches (Reynolds decomposition, $k-\varepsilon$ methods). References to the very rich literature on this subject will be provided.

Reasonable results can also be achieved with "lower costs", by using the integral methods. Although quite old, this approach is still worth being known. It was our intention to initiate the interested reader into these techniques, and a specific paragraph is dedicated to this purpose.

1.3. The outer boundary layers: an analytical approach

1.3.1. *The laminar boundary layer developed by a flat plate in a uniform flow*

Although laminar flows may be perceived as rather academic from a practical point of view, the reasoning developed in regards to this model is highly inciting and formative for the thinking process of fluid mechanics scientists.

Let us consider a fluid with constant physical properties (density μ and kinematic viscosity ν). The potential flow is then obviously a uniform flow of velocity U_e. We shall consider the case of a plane steady flow. An orthonormal frame of reference is attached to the system. The origin is on the leading edge of the plate; x and y axes are parallel and perpendicular to the plate, respectively.

The velocity vector $\vec{V}(u,v)$ verifies the following equations:

Continuity equation:

$$\frac{\partial u}{\partial x}+\frac{\partial v}{\partial y}=0 \qquad [1.8]$$

Impulse equation:

$$u\frac{\partial u}{\partial x}+v\frac{\partial v}{\partial y}=-\frac{1}{\rho}\frac{dp_G}{dx}+\nu\frac{\partial^2 u}{\partial y^2} \qquad [1.9]$$

The term containing $\frac{dp_G}{dx}$ calls for two remarks:

The flow is not rigorously parallel in the boundary layer. In effect, $v(x,y)$ is smaller than $u(x,y)$, but not rigorously null. This makes physical sense: the plate "slows down" the fluid, which is "chased away from the plate" and into the free flow. Nevertheless, it is close enough to a parallel flow to allow us to consider that perpendicularly to these flow lines, and therefore to the plate, the lateral gradient of p_G, $\frac{\partial p_G}{\partial y}$, is null.

A "*full*" d (full derivative) with respect to x is justified here. Similarly, the value of $p_G(x)$ will be to the same as the one observed for the same coordinate x in the free flow.

From a physical perspective, U_e can vary if a wind tunnel of variable cross-section is considered, for example. In the potential flow, where the fluid is deemed perfect, Bernoulli's theorem is verified. Therefore:

$$\frac{d}{dx}\left(\rho\frac{U_e^2}{2}+p_G\right)=0 \qquad [1.10]$$

which offers the possibility to replace $-\frac{1}{\rho}\frac{dp_G}{dx}$ with

$$-\frac{1}{\rho}\frac{dp_G}{dx} = U_e \frac{dU_e}{dx} \quad [1.11]$$

When the potential flow is rigorously uniform, this term is null, a case which will be considered in what follows.

The Blasius problem

Blasius solved the problem of the laminar boundary layer of a fluid with constant physical properties over a flat plate of indefinite width placed in the path of wind of a uniform flow.

The system of equations that relate the two unknowns $u(x,y)$ and $u(x,y)$ becomes

$$\frac{\partial u}{\partial x} + \frac{\partial v}{\partial y} = 0 \quad [1.12]$$

$$u\frac{\partial u}{\partial x} + v\frac{\partial v}{\partial y} = v\frac{\partial^2 u}{\partial y^2} \quad [1.13]$$

The boundary conditions being written

$$y = 0;\; u = v = 0 \quad [1.14.a]$$

$$y \to \infty;\; u \to U_e \quad [1.14.b]$$

It is worth noting that the problem states an asymptotic variation of velocity toward the outer flow "to infinity". In fact, rapid convergence will be noted. A boundary layer thickness is defined by the point in which $u\,(x,\delta) = 0{,}99U_e$. This entails $\delta = \delta(x)$. We shall verify in the following that $\frac{\delta(x)}{x}$ is small.

Blasius showed that the problem becomes self-similar through the definition of a dimensionless composite variable $\eta(x,y)$:

$$\eta = y\sqrt{\frac{U_e}{\nu x}} \quad [1.15]$$

A dimensionless function $f(\eta)$ emerges, which verifies the following differential equation, in which the symbol " ' " signifies one differentiation:

$$f''' + 2ff'' = 0 \quad [1.16]$$

Accompanied by the boundary conditions

$$f(0)=0; f'(0)=0; f'(\infty) \to 1 \quad [1.17]$$

this equation is solved numerically.

The results to be retained are the profile of $f(\eta)$, which gives the profiles of the velocity components u, v and the wall friction expressed by a friction coefficient C_f defined by $C_f = \frac{2\tau_W}{\rho U_e^2}$, where τ_W is the wall shear stress.

We show that the reduced velocity $\frac{u}{U_e}$ is equal to:

$$\frac{u}{U_e} = f'(\eta) \quad [1.18]$$

This leads to the emergence of a self-similarity of velocity profiles.

In particular, for any value of x, the thickness of the boundary layer will be given for

$$f'(\eta) = 0{,}99 \quad [1.19]$$

which is verified for:

$$\eta = 4{,}96 \quad [1.20]$$

$$\eta = \delta \sqrt{\frac{U_e}{\nu x}} = 4,96 \qquad [1.21]$$

$$\delta(x) = 4,96 \sqrt{\frac{\nu x}{U_e}} \qquad [1.22]$$

which is often written as:

$$\frac{\delta(x)}{x} = \frac{4,96}{\sqrt{R_x}} \qquad [1.23]$$

where R_x is the Reynolds number for the x axis:

$$R_x = \frac{U_e x}{\nu} \qquad [1.24]$$

Using the value of $f''(0)$, we obtain:

$$\tau_W = \mu \left.\frac{\partial u}{\partial y}\right|_{y=0} = \mu U_e \left(\left.\frac{df'}{d\eta}\right|_{\eta=0} \right) \frac{\partial \eta}{\partial y} = \mu U_e \sqrt{\frac{U_e}{\nu x}} f''(0) \qquad [1.25]$$

$$f''(0) = 0,332 \qquad [1.26]$$

which finally gives:

$$\frac{C_f}{2} = \frac{\tau_W}{\rho U_e^2} = \frac{0,332}{\sqrt{R_x}} \qquad [1.27]$$

1.3.2. *The turbulent boundary layer*

The turbulent boundary layer, which is more interesting from a more practical perspective, cannot be treated in a similar manner. The analytical approach can only yield mean values of the parameters.

Therefore, an important operation is to calculate these mean values:

We define the mean velocity as:

$$\bar{u}(x,y)=\frac{1}{T}\int_0^T u(t,x,y) \quad [1.28]$$

This time-averaged value theoretically calculated over infinite time is in fact experimentally established over physically long durations (of the order of several minutes).

A Reynolds decomposition ensues:

$$u(x,y)=\bar{u}(x,y)+u' \quad [1.29]$$

Time-averaged u' is null, but not its root mean square; therefore, fluctuations can be calculated by:

$$\frac{1}{T}\int_0^T u'^2(t,x,y) \quad [1.30]$$

Similarly, various correlations will play an important role, such that:

$$\frac{1}{T}\int_0^T u'(t,x,y)v'(t,x,y) \quad [1.31]$$

We shall however see that within the framework of integral methods, similar approaches are applicable to laminar and turbulent layers.

Historically, several approaches have been conducted, generally because of the advancement of metrology. For the boundary layer above a flat plate:

The first studies using the Pitot tube have led to researching mean velocity profiles expressed as power laws (that can be used particularly with integral methods):

$$\frac{\bar{u}(x,y)}{U_e}=\left(\frac{y}{\delta(x)}\right)^n \quad [1.32]$$

where n is a coefficient that varies in principle with Reynolds number. For the usual Reynolds numbers, n takes predominantly the value $\frac{1}{7}$ (1/7th power law).

The development of anemometry (hot-wire anemometry, then Laser Doppler anemometry) facilitated a more detailed rendering of profiles and particularly of the flow structure.

Although the matter is beyond the scope of this initiation manual, it is worth noting that spectral approaches to fluctuations have gained attention. This was made possible by the progressive decrease in sensor response time, from microscopic hot wire to the optical measurement of velocity (Laser Doppler anemometry). We shall mention here the most prominent result, although it will not be used further on.

Variables x^+, y^+ and the dimensionless velocities u^+, v^+ have then been defined by a reference velocity, called friction velocity:

$$U_f = \sqrt{\frac{2\tau_W}{\rho}} = U_e\sqrt{\frac{C_f}{2}} \quad [1.33]$$

$$x^+ = \frac{U_f x}{\nu}; y^+ = \frac{U_f y}{\nu} \quad [1.34]$$

$$u^+ = \frac{\overline{u}}{U_f}; v^+ = \frac{\overline{v}}{U_f} \quad [1.35]$$

The velocity profile has then been divided into three regions.

Film region corresponding to a "viscous" laminar sub-layer with linear velocity profile:

$$u^+ = y^+$$

Wall region comprises most of the boundary layer. This wall law leads to a logarithmic profile with reduced variables:

$$u^+ = \frac{1}{\kappa}\ln y^+ + C \quad [1.36]$$

where κ is von Kármán's constant $\kappa \approx 0.4$.

Region of connection to the potential flow

Using Reynolds decomposition method, the equations for the components of mean velocity can be written as follows:

Using the same calculation technique for the instantaneous equation, the equations can be written in classical form as:

$$\frac{\partial \overline{u}}{\partial x}+\frac{\partial \overline{v}}{\partial y}=0 \tag{1.37}$$

$$\overline{u}\frac{\partial \overline{u}}{\partial x}+\overline{v}\frac{\partial \overline{u}}{\partial y}=-\frac{1}{\rho}\frac{\partial \overline{p}_G}{\partial y}+\nu\frac{\partial^2 \overline{u}}{\partial y^2}-\frac{\partial}{\partial y}\left(\overline{u'v'}\right) \tag{1.38}$$

The system cannot be reduced to these equations of mean velocities.

It is possible to establish an equation for this correlation $\overline{u'v'}$ (this will not be done here). Then, it is noted that the process is divergent: for each differential equation written for a correlation of order *n*, correlations of the order $n+1$ emerge. This is natural. The information lost in the statistical approach (creation of terms in u' and v') requires more than chance to reemerge through simple mathematical manipulations.

Therefore, these equations must be closed, which means that the information must be fed back in through a supplementary relation. This closure can be performed at various n levels.

The best known closure, and one of the simplest, is of the gradient type, such as that proposed by Boussinesq:

$$\overline{u'v'}=-\varepsilon_v\frac{\partial \overline{u}}{\partial y} \tag{1.39}$$

where ε_v is turbulent viscosity. Let us note that ε_v has the dimension of a kinematic viscosity.

The significance of the term $\overline{u'v'}$ in the equations is that of shear stress, and hence it is called "Reynolds stress". It contains the mechanism of "turbulent friction" illustrated by the Boussinesq closure.

The viscosity of a fluid is due to the transfer of momentum between the layers of fluid that are in relative translation by molecular agitation. Similarly, eddies of the turbulent flow carry momentum through the velocity gradient. In this case, the scale of the phenomenon is the scale of the eddy. This was used by Prandtl in his "mixing length" theory.

Moreover, the values obtained for turbulent viscosity are much higher; "molecular" viscosity will subsequently be often neglected (with the exception of the laminar sub-layer).

In the field of inner boundary layers, the problem will most often be solved empirically. Since the established regime is not predominant, Moody charts and a Darcy type of approach will be used to determine frictions.

The values that are often retained for the thickness of the boundary layer and the friction over a flat plate are deduced from the value of the friction coefficient:

$$\frac{\delta}{x} = \frac{0,382}{R_x^{0.2}} \quad [1.40]$$

$$C_f = \frac{0,0594}{R_x^{0,2}} \quad [1.41]$$

where $R_x = \dfrac{U_e\, x}{\nu}$ is the Reynolds number for the x axis.

Many semi-empirical methods have been tried in the past for the treatment of turbulent flows. The predominant method currently applied is the numerical approach, which is commonly used in many industrial applications. On this subject, the reader is invited to see the chapters dedicated to numerical processing.

Several problems pertaining to this class of approaches will be provided below. In particular, an introduction to the treatment of non-Newtonian fluids will be offered.

1.4. Examples of analytical approach: outer flows

Comprehensive works have been dedicated to the subject of boundary layer theory. For an introduction on the subject, it is worth presenting the reader with some rather atypical and original problems. Therefore, this section focuses on unsteady boundary layers (two classical problems are proposed) and an introduction to non-Newtonian fluids. As we shall see, when dealing with boundary layers or tubes, it is worth considering the rheological specificity of certain fluids in a simple manner, without reducing it to a Newtonian fluid hypothesis, which may sometimes prove too reductive.

For an approach of unsteady flows, a higher dimension should be considered. This may prove a complex issue, as the dimension of the problem becomes $n+1$, where n is the number of space dimensions considered.

In-depth study of unsteady flows is not the object of this book. Several classical examples that are physically useful will be presented here for the dimension n reduced to 1. These solutions appear as welcome simplifications of more complex problems of unsteady boundary layers. We shall establish in passing an interesting connection between an unsteady setting and a steady boundary layer.

As far as non-Newtonian fluids are concerned, we shall consider power law fluids or Eyring fluids.

EXAMPLE 1.1 (Unsteady boundary layer. Stokes' first problem).–

Let us consider the uniform flow of velocity U of an incompressible fluid of density ρ and dynamic viscosity μ. Kinematic viscosity of this fluid will be expressed as $\nu = \dfrac{\mu}{\rho}$. The flow lines are considered horizontal.

At time $t=0$, a flat plate of infinite extension is placed in this fluid, parallel to the flow lines. Given the symmetry of the problem relative to the plate, only the flow over the upper face will be studied. The flow lines are also considered to remain parallel to the wall at any time subsequent to $t=0$.

The axis perpendicular to the wall will be denoted by Oy, the origin of the coordinates of which being located on this wall.

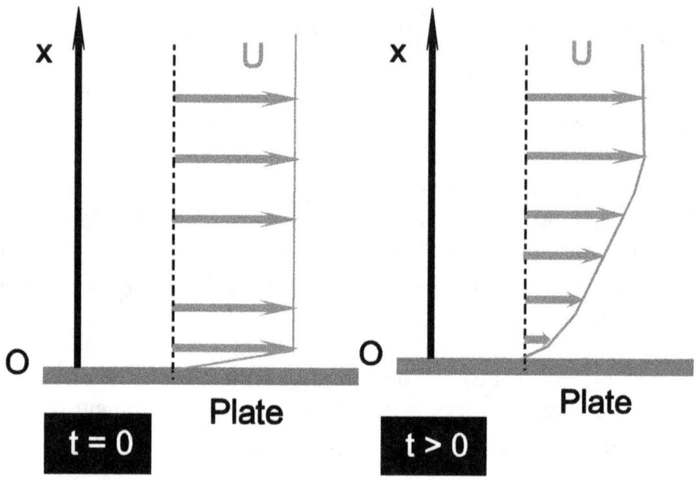

Figure 1.1. *Stokes' first problem. Representation of the velocity profile at two moments of time*

1) Show that velocity assumes the form $\vec{V}(y,t)$ in Eulerian description.

Write the differential equation which is satisfied by $u(y,t)$ and the boundary conditions of the problem for $t > 0$.

2) What does the form of this equation remind you of?

3) Find the expression of $u(y,t)$. It is useful to introduce the variable η:

$$\eta = \frac{y}{\sqrt{4\nu t}} \quad [1.42]$$

4) Let us now consider the boundary layer developed by a uniform flow of outer velocity U_e over a semi-infinite plate placed under the path of wind. By modeling a section of thickness dx perpendicular to the plate, whose displacement starting from the leading edge is observed, and which plays the role of the previous flow, finds an approximate shape of the velocity $u(y)$ profile for a given abscissa x, which will be an approximation of $u(x,y)$ in the boundary layer.

Deduce from this an approximate formula for the thickness of the dynamic boundary layer $\delta(x)$. Compare this form with the exact result of the Blasius analysis.

Solution:

1) The flow is plane.

The velocity $\vec{V}$ has only one component u. For an incompressible fluid, the continuity equation is:

$$\frac{\partial u}{\partial x}+\frac{\partial v}{\partial y}=0 \qquad [1.43]$$

As v is null, $\frac{\partial u}{\partial x}=0$ and u can only be a function of y and t, since the flow is unsteady. Then, the projection on Ox of the equation of (dynamic) impulse for a plane flow of Newtonian fluid can be written as:

$$\rho\left(\frac{\partial u}{\partial t}+u\frac{\partial u}{\partial x}+v\frac{\partial u}{\partial y}\right)=-\frac{1}{\rho}\frac{\partial p}{\partial x}+\mu\frac{\partial^2 u}{\partial y^2} \qquad [1.44]$$

where v is null, and so is the term containing pressure. Introducing kinematic viscosity, the above relation becomes:

$$\frac{\partial u}{\partial t}=\nu\frac{\partial^2 u}{\partial y^2} \qquad [1.45]$$

Two boundary conditions are associated to this equation:

$$t>0 \ y=0; u=0 \qquad [1.46.a]$$

$$t>0 \ y\rightarrow\infty; u\rightarrow U \qquad [1.46.b]$$

2) This equation is formally identical to the equation of heat under one-dimensional and unsteady regime, in which ν plays the role of thermal diffusivity a.

Although it is beyond the scope of this work, it is worth noting that the symmetry of the roles of ν and a is reflected by the impulse and energy equations written for fluids with constant physical properties. ν and a have the same dimension, L^2T^{-1}, and their ratio defines one of the most important numbers in the theory of thermal convection, the Prandtl number $P_r = \frac{\nu}{a}$

3) To solve this problem, the approach taken in the theory of thermal conduction will be referred to as a model. The following variable is introduced:

$$\eta = \frac{y}{\sqrt{4\nu t}} \qquad [1.47]$$

This variable is the analogue of $\eta = \frac{y}{\sqrt{4ax}}$ in the theory of unsteady state heat conduction (in this case, a is thermal diffusivity).

Differentiations are performed as follows:

$$\frac{\partial \eta}{\partial t} = -\frac{y}{\sqrt{4\nu}} \frac{1}{2t^{3/2}} = -\frac{\eta}{2t} \qquad [1.48]$$

$$\frac{\partial \eta}{\partial y} = \frac{1}{\sqrt{4\nu t}}; \frac{\partial^2 \eta}{\partial y^2} = 0 \qquad [1.49]$$

$$\frac{\partial u}{\partial t} = \frac{\partial u}{\partial \eta}\frac{\partial \eta}{\partial t}; \frac{\partial u}{\partial y} = \frac{\partial u}{\partial \eta}\frac{\partial \eta}{\partial y}; \frac{\partial^2 u}{\partial y^2} = \frac{\partial u}{\partial \eta}\frac{\partial^2 \eta}{\partial y^2} + \frac{\partial^2 u}{\partial \eta^2}\left(\frac{\partial \eta}{\partial y}\right)^2 \qquad [1.50]$$

$$\frac{\partial u}{\partial \eta}\frac{\eta}{2t} = -\frac{\partial^2 u}{\partial \eta^2}\frac{1}{4\nu t} \qquad [1.51]$$

Time is eliminated from this equation and u appears as a unique function of η, which satisfies the equation accompanied by its boundary conditions:

$$\frac{d^2 u}{d\eta^2} = -2\eta \frac{\partial u}{\partial \eta} \qquad [1.52]$$

$$\eta = 0;\ u = 0;\ \eta \to \infty;\ u \to U \qquad [1.53]$$

The solution yields immediately:

$$\frac{\frac{d^2 u}{d\eta^2}}{\frac{\partial u}{\partial \eta}} = -2\eta \tag{1.54}$$

$$Ln\frac{\partial u}{\partial \eta} = -\eta^2 + Ln\, C_1 \tag{1.55}$$

$$\frac{\partial u}{\partial \eta} = C_1 \exp - \eta^2 \tag{1.56}$$

$$u = \int C_1 \exp - \xi^2 + C_2 \tag{1.57}$$

The boundary conditions yield, knowing that:

$$\int_0^\infty \exp - \xi^2 = \frac{\sqrt{\pi}}{2} \tag{1.58}$$

$$u = C_1 \int_0^\eta \exp - \xi^2 \, d\xi \tag{1.59}$$

$$C_1 = \frac{2U}{\sqrt{\pi}} \tag{1.60}$$

The error function can be identified in [1.59], up to a coefficient:

$$erf(\eta) = \frac{2}{\sqrt{\pi}} \int_0^\infty \exp - \xi^2 \, d\xi \tag{1.61}$$

$u(y,t)$ is finally expressed as:

$$u = U\, erf \frac{y}{\sqrt{4\nu t}} \tag{1.62}$$

4) The approximation mechanism proposed in this section should be properly understood. It amounts to applying to a thin section a result established for an indefinite medium. In the proposed pattern, which follows the section in its motion, x is identified as Ut.

Then, $u(t,y)$ is identical to $u(x,y)$ and the field of velocity within the framework of this approximation can be written as:

$$\frac{u(x,y)}{U} = erf\left(y\sqrt{\frac{U}{4\nu x}}\right) \quad [1.63]$$

The conventional limit of this boundary layer is reached for $u = 0.99U$, which is an argument of the *erf* function equal to 1.82. This yields:

$$y\sqrt{\frac{U}{4\nu\delta}} = 1,82 \quad [1.64]$$

$$\frac{\delta}{x} = 3,64\sqrt{\frac{\nu x}{U}} = \frac{3,64}{R_x} \quad [1.65]$$

Reynolds number for the x axis $R_x = \sqrt{\frac{Ux}{\nu}}$ has been introduced. This result is comparable to that of the Blasius analysis:

$$\frac{\delta}{x} = \frac{4,95}{R_x} \quad [1.66]$$

The error is 26%, for quite an elementary and rapid solving.

EXAMPLE 1.2 (Stokes' second problem).–

A horizontal flat plate of length L and indefinite width oscillates parallel to itself. A frame of reference *Oxy* is defined (this is obviously a two-dimensional phenomenon). *Ox* is parallel to the plate along its axis of movement and *Oy* is perpendicular to the plate.

The velocity of the plate is given by $U(t) = U_0 \cos \omega t$

This plate is immersed in an incompressible fluid of density ρ and dynamic viscosity μ. The fluid region set in motion is considered thin relative to the longitudinal dimension L of the plate (which validates the hypothesis of the indefinite plate). It can be readily admitted that fluid velocity above the wall has only a horizontal component.

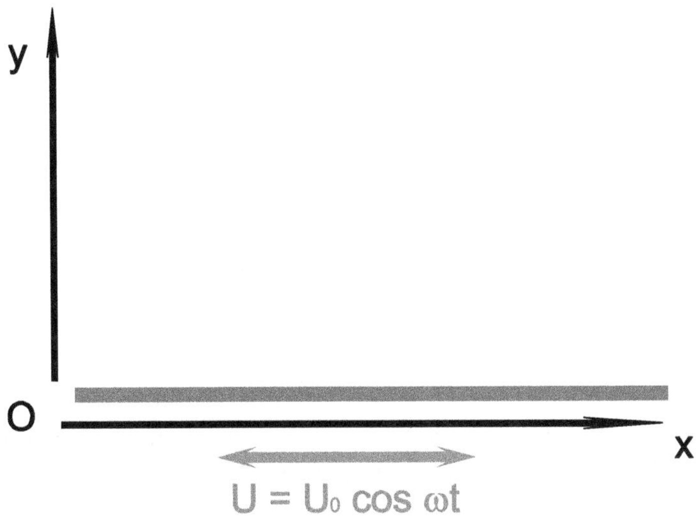

Figure 1.2. *Stokes' second problem*

1) The frequency of vibration f is high enough for the Strouhal number to be considered very large, which will be defined as:

$$S = \frac{Lf}{U_0} \qquad [1.67]$$

where L becomes a length-scale characteristic to the problem.

Considering the orders of magnitude for the problem of various terms of the impulse equation, find its very simplified form.

Quickly show that velocity above the plate has the form $u(y,t)$.

Write the equation that is satisfied by u.

Write the boundary conditions for y.

NOTE.– The problem is considered in the context of the plate vibrating for a sufficiently long time. The flow is therefore in "steady-state regime".

All unsteady problem of this type involves a transient solution (from the moment the plate is set in motion) and a "steady-state" solution. This latter type of solution is of interest here, given the previous hypothesis.

2) Find the form of $u(y,t)$

Finding this solution will be significantly facilitated by using complex numbers. $u(y,t)$ will then be the real part of the complex solution, which is written as:

$$u_{Complex} = \Phi(y)e^{i\omega t} \quad [1.68]$$

Let us recall the identity:

$$\sqrt{i} = \frac{1+i}{\sqrt{2}} \quad [1.69]$$

3) Show that there is a phase difference $\phi(y)$ between the velocities of the two planes of fluid parallel to the wall. What is the distance δ between two in-phase planes?

What is the distance l from the wall above which the velocity amplitude Φ will be one hundredth of U_0?

4) Numerical application.

The plate oscillates in water with $f = 500\,Hz$. We know that $\rho = 1000\,kg.m^{-3}$ and $\mu = 10^{-3}\,Pl$. Find the values of δ and $S = l$.

Show that this device can be used to measure viscosity. What type of viscosity is directly measured? If δ is measurable up to 10%, what is the accuracy for ν?

Solution:

1) The flow is plane

The velocity $\vec{V}$ has only one component u.

For an incompressible fluid, the continuity equation is:

$$\frac{\partial u}{\partial x} + \frac{\partial v}{\partial y} = 0 \quad [1.70]$$

as v is null:

$$\frac{\partial u}{\partial x} = 0 \quad [1.71]$$

and u can only be a function of y and t, since the flow is unsteady.

Then, the projection on Ox of the equation of (dynamic) impulse for a plane flow of Newtonian fluid can be written as:

$$\rho\left(\frac{\partial u}{\partial t}+u\frac{\partial u}{\partial x}+v\frac{\partial u}{\partial y}\right)=-\frac{1}{\rho}\frac{\partial p}{\partial x}+\mu\frac{\partial^2 u}{\partial y^2} \quad [1.72]$$

v is null, and so is the term containing pressure. Introducing kinematic viscosity, the above relation becomes:

$$\frac{\partial u}{\partial t}=\nu\frac{\partial^2 u}{\partial y^2} \quad [1.73]$$

Two boundary conditions are associated to this equation:

$$t>0\ y=0;\ u=U_0\cos\omega t \quad [1.74]$$

$$t>0\ y\to\infty;\ u\to 0 \quad [1.75]$$

2) The context is that of "steady-state regime" (out of transient regime), and we search for u under an associated complex form $u_{Complex}$:

$$u_{Complex}=\Phi(y)e^{i\omega t} \quad [1.76]$$

In order to satisfy the differential equation, $\Phi(y)$ has to verify:

$$i\omega\,\Phi(y)e^{i\omega t}=\nu\frac{d^2\Phi}{d\,y^2}e^{i\omega t} \quad [1.77]$$

$$\frac{d^2\Phi}{dy^2}-\frac{i\omega}{\nu}\Phi=0 \quad [1.78]$$

A classical equation with the general solution:

$$\Phi(y)=A\exp\sqrt{\frac{i\omega}{\nu}}y+B\exp-\sqrt{\frac{i\omega}{\nu}}y \quad [1.79]$$

$$\Phi(y)=A\exp\frac{1+i}{\sqrt{2}}\sqrt{\frac{\omega}{\nu}}y+B\exp-\frac{1+i}{\sqrt{2}}\sqrt{\frac{\omega}{\nu}}y \quad [1.80]$$

$$\Phi(y) = A\exp\left(\sqrt{\frac{\omega}{2\nu}}y\right)\exp i\left(\omega t + \sqrt{\frac{\omega}{2\nu}}y\right) + B\exp\left(-\sqrt{\frac{\omega}{2\nu}}\right)\exp i\left(\omega t - \sqrt{\frac{\omega}{2\nu}}y\right) \quad [1.81]$$

$\Phi(y)$ has to tend to 0 at infinity; therefore, A is null.

$u_{Complex} = \Phi(y)e^{i\omega t}$ can then be written as:

$$u_{Complex} = B\exp\left(-\sqrt{\frac{\omega}{2\nu}}y\right)\exp i\left(\omega t - \sqrt{\frac{\omega}{2\nu}}y\right) \quad [1.82]$$

$$u_{Complex} = B\exp\left(-\sqrt{\frac{\omega}{2\nu}}y\right)\left[\cos\left(\omega t - \sqrt{\frac{\omega}{2\nu}}y\right) + i\sin\left(\omega t - \sqrt{\frac{\omega}{2\nu}}y\right)\right] \quad [1.83]$$

Taking the real part:

$$u_{Complex} = B\exp\left(-\sqrt{\frac{\omega}{2\nu}}y\right)\left[\cos\left(\omega t - \sqrt{\frac{\omega}{2\nu}}y\right)\right] \quad [1.84]$$

The wall condition requires that $B = U_0$

$$u_{Complex} = U_0\exp\left(-\sqrt{\frac{\omega}{2\nu}}y\right)\left[\cos\left(\omega t - \sqrt{\frac{\omega}{2\nu}}y\right)\right] \quad [1.85]$$

It can be noted that the plate imposes a disturbance that decreases exponentially with the distance to the plate.

3) Moreover, all the planes that are parallel to the plate vibrate at the same frequency, but are out of phase. The phase difference between two planes separated by y is:

$$\phi = \sqrt{\frac{\omega}{2\nu}}y \quad [1.86]$$

Two planes in phase are separated by δ:

$$\phi = 2\pi = \sqrt{\frac{\omega}{2\nu}}\delta \quad [1.87]$$

$$\delta = \sqrt{\frac{8\pi^2 \nu}{\omega}} \qquad [1.88]$$

The velocity equals one hundredth of U_0 for:

$$\exp\left(-\sqrt{\frac{\omega}{2\nu}}l\right) = 10^{-2}; \quad \sqrt{\frac{\omega}{2\nu}}l = Ln 10^{-2} = 4,6 \qquad [1.89]$$

$$l = 4,6\sqrt{\frac{2\nu}{\omega}} \qquad [1.90]$$

4) Numerical application

$f = 500\, Hz$; $\omega = 1000\pi; \nu = 10^{-6}\, m.s^{-2}$

Let:

$$\delta = 1,59.10^{-4}\, m = 159\, \mu m \qquad [1.91]$$

$$l = 4,6\sqrt{\frac{2\nu}{\omega}} = 1,1.10^{-4}\, m = 110\, \mu m \qquad [1.92]$$

For a relative error of 10% on δ, the accuracy of the measurement of kinematic viscosity is:

$$\frac{\Delta\nu}{\nu} = 2\frac{\Delta\delta}{\delta} \qquad [1.93]$$

The error is 20%.

1.5. Examples of analytical approach: inner flows

EXAMPLE 1.3 (Couette flow. Newtonian and non-Newtonian perfect fluids).–

The Couette flow is generated between two horizontal, parallel, flat plates of indefinite dimensions. One of the plates moves in parallel to itself with a velocity U_0, while the other plate is at rest. The two plates are separated by a fluid of thickness e.

This physical situation models what happens in the gap between the two cylinders of a Couette viscometer, if we consider that the radius of the cylinders is very large compared with their radius difference.

For this reason, the pressure will be considered unchanged in the direction parallel to the plates.

A frame of reference of axes Oxy is attached to the system (the flow is obviously plane), with Ox parallel to the plate. The origin O is located on the lower plate, which moves to the right with velocity U_0.

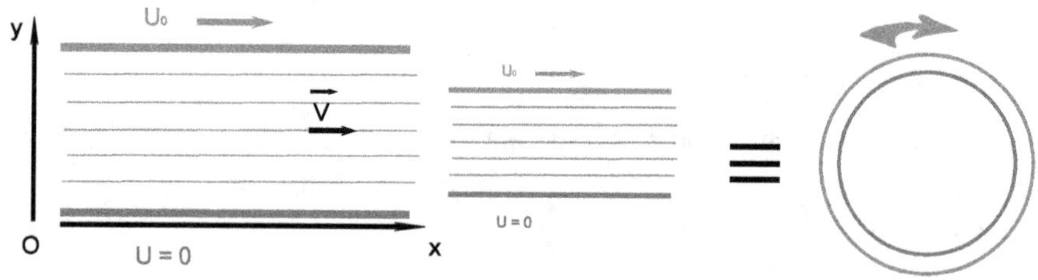

Figure 1.3. *Couette flow. On the left: Couette flow between two plates. On the right: application of this model as an approximation to Couette viscometer*

1) A perfect fluid is placed between the two plates: What happens?

2) Then an incompressible fluid of density ρ and dynamic viscosity μ is placed between the plates. Find the expression of the components u and v of the velocity field $\vec{V}$. Find the expression of the shear stress τ_W on each plate.

3) An Ostwald–de Waele fluid is placed between the plates. Consider question 2 for this type of fluid.

4) Resume question 2 for an Eyring fluid.

REMINDER.–

Ostwald and de Waele showed that for certain non-Newtonian fluids, the relationship between shear stress and a one-dimensional velocity gradient $\frac{du}{dy}$ can be expressed as:

$$\tau = k\left(\frac{du}{dy}\right)^n \qquad [1.94]$$

where τ is the shear stress, k is the "coefficient of viscosity" (or compacity, which does not have the dimension of a dynamic viscosity) and n is a coefficient.

If $n<1$, the fluid is called pseudo-plastic; if $n>1$, it is called dilatant.

Based on the works of Eyring, for certain non-Newtonian fluids, the following relation between shear stress $n<1$ and shear strain rate $\frac{du}{dy}$ can be written as:

$$\tau = A\, sh^{-1}\left(\tau_0 \frac{du}{dy}\right) \qquad [1.95]$$

where τ_0 is a constant and $sh^{-1}\,x$ (or $\arg shx$) is the inverse function of hyperbolic sine shx.

The Powell–Eyring general law is more complex:

$$\tau = \mu \frac{du}{dy} \qquad [1.96]$$

where:

$$\mu - \mu_\infty = \frac{(\mu_0 - \mu_\infty)}{\tau_0 \frac{du}{dy}} sh^{-1}\left(\tau_0 \frac{du}{dy}\right) \qquad [1.97]$$

The form provided here corresponds to a pseudo-plastic fluid, for which μ_∞ is very low. μ_∞ and τ_0 are constants. In this case, μ_∞ will be neglected.

Solution:

1) The perfect fluid hypothesis

By definition, in a perfect fluid, the tangential component of surface forces is null. Therefore, the fluid does not exert any tangential force on the walls, so there is no braking or drive. Similarly, following Newton's "third" law, referring to action and reaction, the plate exerts only normal forces (pressure) on the fluid, and therefore will exert no drive or braking action on the fluid. The plates will "slip perfectly" over the fluid, which remains perfectly still. Referring to this particular point, it is worth noting that the proposed hypothesis is contradicted by experience. This contradiction makes the object of the so-called "d'Alembert's paradox".

NOTE.– Let us recall that too many students similarly forget to consider Newton's "second" law, also known as the fundamental principle of dynamics. The quotation marks are a reminder that Newton's "first" law is in fact an inertia principle that had already been formulated by Galileo, a "principle" that can be proven starting from Newton's "second" law.

2) Newtonian fluid

Although not explicitly provided in the text, each time we introduce a fluid of density ρ and dynamic viscosity μ, the assumption being made is that the fluid is Newtonian (in general, a non-Newtonian fluid is defined based on coefficients, and not on viscosity; to be extremely rigorous, a formula is provided for the viscosity, which relates it to stresses). Moreover, the other questions of the problem leave no room for ambiguity.

The fluid being incompressible, we readily proceed to defining its kinematic viscosity ν, which appears in the equation of impulse: $\nu = \dfrac{\mu}{\rho}$

Kinematic remark: For plates of indefinite dimensions, it is not difficult to see that the flow is parallel (obviously non-uniform).

Therefore, the component v of the flow will necessarily be null.

In order to get information, let us use the continuity principle and the second law of dynamics (impulse equation).

The continuity equation gives:

$$\frac{\partial u}{\partial x} + \frac{\partial v}{\partial y} = 0 \qquad [1.98]$$

$$v = 0; \frac{\partial u}{\partial x} = 0; u = u(y) \qquad [1.99]$$

There is only one component of velocity left, $u(y)$, exclusively dependent on y, which will satisfy the impulse equation written for an incompressible Newtonian fluid (two-dimensional Navier–Stokes equation):

$$\frac{\partial u}{\partial t} + u\frac{\partial u}{\partial x} + v\frac{\partial u}{\partial y} = -\frac{1}{\rho}\frac{\partial p}{\partial x} + 2\nu\frac{\partial^2 u}{\partial x^2} + \nu\left[\frac{\partial^2 u}{\partial y^2} + \frac{\partial v}{\partial x \partial y}\right] + F_{Vx} \qquad [1.100]$$

$$\frac{\partial v}{\partial t}+u\frac{\partial v}{\partial x}+v\frac{\partial v}{\partial y}=-\frac{1}{\rho}\frac{\partial p}{\partial y}+\nu\left(\frac{\partial^2 u}{\partial x\partial y}+\frac{\partial^2 v}{\partial x^2}\right)+\frac{\partial}{\partial y}2\nu\left(\frac{\partial^2 u}{\partial y^2}\right)+F_{Vy} \quad [1.101]$$

When projecting the equation on Oy, all the terms containing velocities will be null. The volume forces are reduced to the gravity term, which, given that Oy is a vertical coordinate, can be written as: $F_{Vy}=-g$

This equation comes down to:

$$-\frac{1}{\rho}\frac{\partial p}{\partial y}-g=0 \quad [1.102]$$

Introducing $p_G=p+\rho y$, we get:

$$-\frac{1}{\rho}\frac{\partial p+\rho y}{\partial y}=-\frac{\partial p_G}{\partial y}=0 \quad [1.103]$$

By projecting the equation on Ox, we obtain:

$$\frac{d^2 u}{dy^2}=0 \quad [1.104]$$

Here we are using full derivatives, since u depends only on y.

This relation also expresses that:

$$\frac{d}{dy}\tau_w=0 \quad [1.105]$$

Since in particular (see the text):

$$\frac{\partial p}{\partial x}=0 \quad [1.106]$$

u is readily deduced from the simple differential equation. Two boundary conditions are needed and they readily obtain:

$$\frac{d^2 u}{dy^2}=0 \quad [1.107]$$

$$y=0\Rightarrow u=0;\ y=e\Rightarrow u=U_0 \quad [1.108]$$

The solution is trivial. The fact that the second derivative is null indicates that u is a linear function. The application of boundary conditions yields:

$$u(y)=U_0\frac{y}{e} \quad [1.109]$$

For a Newtonian fluid, shear stress is expressed as:

$$\tau_W=\mu\frac{du}{dy}=\mu\frac{U_0}{e} \quad [1.110]$$

It has the same "absolute value" on each of the plates.

3) The Ostwald–de Waele fluid

Given the flow structure and the results of question 2, we have:

$$u=u(y);\, v=0 \quad [1.111]$$

$$\frac{d}{dy}\tau=0 \quad [1.112]$$

The stress can be written as:

$$\tau=k\left(\frac{du}{dy}\right)^n \quad [1.113]$$

The equation to be solved and its boundary conditions become:

$$\frac{d}{dy}k\left(\frac{du}{dy}\right)^n=0 \quad [1.114]$$

$$y=0\Rightarrow u=0;\, y=e\Rightarrow u=U_0 \quad [1.115]$$

The resolution is simple:

$$k\left(\frac{du}{dy}\right)^n=C_1 \quad [1.116]$$

$$\frac{du}{dy}=\left(\frac{C_1}{k}\right)^{1/n}=C_1' \quad [1.117]$$

After integration:

$$u = C_1' y + C'_2 \qquad [1.118]$$

where C_1, C_1' and C_2' are constants.

In order to satisfy the boundary conditions:

$$u(y) = U_0 \frac{y}{e} \qquad [1.119]$$

which is the same result as for the Newtonian fluid.

Similarly, the wall stresses will be:

$$\tau_W = \tau = k\left(\frac{du}{dy}\right)^n = k\left(\frac{U_e}{e}\right)^n \qquad [1.120]$$

4) Eyring fluid

Given the flow structure and the results of question 2, we have:

$$u = u(y); v = 0 \qquad [1.121]$$

$$\frac{d}{dy}\tau = 0 \qquad [1.122]$$

The stress has the form:

$$\tau = A\, sh^{-1}\left(\tau_0 \frac{du}{dy}\right) \qquad [1.123]$$

where

$$A = \frac{(\mu_0 - \mu_\infty)}{\tau_0} \qquad [1.124]$$

In this case, $\mu_\infty << \mu$

The equation to be solved and its boundary conditions become:

$$\frac{d}{dy} A\, sh^{-1}\left(\tau_0 \frac{du}{dy}\right) = 0 \qquad [1.125]$$

$$y = 0 \Rightarrow u = 0;\ y = e \Rightarrow u = U_0 \qquad [1.126]$$

A first integration yields:

$$sh^{-1}\left(\tau_0 \frac{du}{dy}\right) = C_1 \qquad [1.127]$$

which is equivalent to

$$\frac{du}{dy} = \frac{sh C_1}{\tau_0} = C'_1 \qquad [1.128]$$

$$y = 0\ u = 0;\ y = e\ u = U_0 \qquad [1.129]$$

The boundary conditions determine C'_1, and this refers us to the two previous problems:

$$u(y) = U_0 \frac{y}{e} \qquad [1.130]$$

$$\tau_W = \tau = A\, sh^{-1}\left(\tau_0 \frac{du}{dy}\right) = \tau = A\, sh^{-1}\left(\tau_0 \frac{U_e}{e}\right) \qquad [1.131]$$

In a Couette viscometer, U_e is determined by the velocity of the cylinder and the shear strain rate is proportional to the $\frac{U_e}{e}$ ratio. Even though the velocity profile is, in theory, linear (which allows for a direct determination of the shear strain rate), in the case of a Newtonian fluid, the Couette viscometer will demonstrate the expected linearity between stress and shear strain rate, while for de Waele and Eyring fluids this relation is nonlinear.

EXAMPLE 1.4 (Non-Newtonian flow in a pipe).–

We have already recalled that Ostwald and de Waele have shown that for certain non-Newtonian fluids, the relation between the shear stress and a one-dimensional velocity gradient $\frac{du}{dy}$ can be expressed as:

$$\tau = k\left(\frac{du}{dy}\right)^n \qquad [1.132]$$

where τ is the shear stress, k is a "coefficient of viscosity" (which does not have the dimension of a dynamic viscosity) and n is a coefficient.

This problem is restricted to fluids for which:

$$(-1)^n = -1 \qquad [1.133]$$

The objective here is to study the flow of such fluid as laminar flow through a horizontal tube of radius R.

The following hypotheses are accepted:

The flow is steady

The regime is steady-state:

The paths of the particles of fluid are lines parallel to the axis of the cylinder.

The flow is of axial revolution.

The fluid does not slip on the solid wall.

The shear stress retains a finite value.

A cylindrical coordinate system (x, r, θ) is used. The stagnation pressure on a plane normal to the axis of the tube is denoted by p_G.

Let us consider an elementary annular space between the two cylinders of radii r and r + dr, limited by two planes normal to the axis of the tube, separated by dx.

1) Find and write the expressions of the forces exerted on this annular space.

2) Write Euler's theorem for this space.

3) Deduce the differential equation relating $u(r)$ and $p_G(x)$ from it. Solve this equation. As a particular case, show that $\frac{dp_G}{dx}$ is a constant, which will be denoted by $-K$.

4) Find the expression of $u(r)$. In order to correctly make explicit the term $\left|\frac{du}{dy}\right|$, we shall consider that the profile of the velocity $u(r)$ is a monotone decreasing function of r. This hypothesis will subsequently be validated.

4.1) Find the expression of $u(r)$ as a function of k, n, R and K. Verify the validity of the hypothesis for $u(r)$.

4.2) Find the expression of the flow velocity V_q.

4.3) Then, express K as a function of the volume flow rate Q_V, as well as of k, n and R.

5) The head loss is to be expressed according to Darcy

Find the relation between p_G and Q_V. Write it as a Darcy-type expression, where J is the head loss per unit length and ψ is a coefficient of head loss per unit length, which can be expressed as:

$$J = \psi \frac{V_q^2}{2gD} \quad [1.134]$$

6) For non-Newtonian fluids, an equivalent Reynolds number is often introduced, with the following definition:

$$R_{en} = 8^{1-n} \left(\frac{4n}{3n+1} \right)^n \frac{\rho V_q^{2-n} D^n}{k} \quad [1.135]$$

Show that R_{en} is dimensionless ($D = 2R$ is the diameter of the tube).

What is the purpose of this definition?

How are these expressions written when $n = 1$?

What is the physical case they relate to?

Solution:

1) This is a case of established flow. The velocity $\vec{V}$ is reduced to its axial component u.

Let us project these forces on Ox. The only forces with non-null projection are the surface forces: viscosity forces and pressure forces.

The resultant of pressure and gravity forces is:

$$-2\pi r\, dr \frac{dp_G}{dx} dx \qquad [1.136]$$

We are using here p_G, which takes into account the effect of gravity forces on the local pressure. The balance of pressure forces is then applied to opposite surface elements of the same elevation. The result thus obtained can be extended to a tube that is inclined toward the horizontal direction. Let us calculate the viscosity forces.

On the inner cylinder of radius r, they equal:

$$-2\pi r\, dx\, \tau = -2\pi r\, dx\, k\left(\frac{du}{dy}\right)^n \qquad [1.137]$$

On the outer cylinder (radius $r + d\,r$), they equal:

$$-2\pi r\, dx\, k\left(\frac{du}{dy}\right)^n + \frac{d}{dr}\left(-2\pi r\, dx\, k\left(\frac{du}{dy}\right)^n\right) dr \qquad [1.138]$$

Only fluids in which $(-1)^n = -1$ are considered. The sign to be attached to each of these expressions is the same, irrespective of the local direction of variation of u with y.

The projection on Ox of the final balance of forces is:

$$-2\pi r\, dr \frac{dp_G}{dx} dx - 2\pi r\, dx\, k\left(\frac{du}{dy}\right)^n + \frac{d}{dr}\left(-2\pi r\, dx\, k\left(\frac{du}{dy}\right)^n\right) dr$$

$$= -2\pi r\, dr \frac{dp_G}{dx} dx - \frac{d}{dr}\left(2\pi r\, dx\, k\left(\frac{du}{dy}\right)^n\right) dr \qquad [1.139]$$

2) Let us write Euler's theorem for a volume of fluid limited by:

two cylinders of radii r and $r+dr$ and two planes perpendicular to *Ozr* (therefore vertical) separated by a distance *dx*.

Each particle of fluid of this element moves with constant velocity (vector quantity defined by direction, orientation and hourly speed). Therefore, its acceleration is null. The momentum entering the reference volume is equal to the momentum leaving it, thereby making the resultant of forces applied to it null.

$$-2\pi r\, dr \frac{dp_G}{dx} dx - \frac{d}{dr}\left(2\pi r\, dx\, k\left(\frac{du}{dy}\right)^n\right) dr = 0 \qquad [1.140]$$

3) After simplification by constant terms and several rearrangements, the sought-for differential equation is finally found:

$$\frac{1}{r}\frac{d}{dr}\left(r\,k\left(\frac{du}{dy}\right)^n\right) = \frac{dp_G}{dx} \qquad [1.141]$$

The solving of this system draws on Poiseuille's theory:

The equation equalizes a term exclusively dependent on r and the one exclusively dependent on z. These terms cannot be equal unless they are constant. Let us denote this constant K. It has been chosen to be positive and it can be predicted that $\frac{dp_G}{dx}$ will be negative:

$$\frac{1}{r}\frac{d}{dr}\left(rk\left(\frac{du}{dy}\right)^n\right) = -K \qquad [1.142]$$

$$\frac{dp_G}{dx} = -K \qquad [1.143]$$

4) Resolving the equation for $u(r)$

4.1) The equation for u is a second-order equation and it calls for the following two boundary conditions:

A no-slip condition in $r = R$: $r = R; u = 0$

The expression of volume flow rate: $Q_v = \int_0^R u(r)\, 2\pi r dr$

Let us find the general form of $u(r)$:

$$\frac{1}{r}\frac{d}{dr}\left(rk\left(\frac{du}{dy}\right)^n \right) = -K \qquad [1.144]$$

$$\frac{d}{dr}\left(rk\left(\frac{du}{dy}\right)^n \right) = -K\,r \qquad [1.145]$$

$$rk\left(\frac{du}{dy}\right)^n = -K\frac{r^2}{2} + C_1 \qquad [1.146]$$

$$\frac{du}{dy} = \left(-\frac{K}{k}\frac{r}{2} + \frac{C_1}{r} \right)^{\frac{1}{n}} \qquad [1.147]$$

Since the derivative should obviously be finite when $r = 0$ (continuity of profile), C_1 is nullified.

$$u = \left(-\frac{K}{2k} \right)^{\frac{1}{n}} \frac{n}{n+1} r^{\frac{n+1}{n}} + C_2 \qquad [1.148]$$

Given that $(-1)^n = -1$, the no-slip condition leads to:

$$C_2 = \left(\frac{K}{2k} \right)^{\frac{1}{n}} \frac{n}{n+1} R^{\frac{n+1}{n}} \qquad [1.149]$$

$u(r)$ is written as:

$$u = \left(-\frac{K}{2k} \right)^{\frac{1}{n}} \frac{n}{n+1} \left(R^{\frac{n+1}{n}} - r^{\frac{n+1}{n}} \right) \qquad [1.150]$$

or in a more canonical form:

$$u=\left(-\frac{K}{2k}\right)^{\frac{1}{n}}\frac{n}{n+1}R^{\frac{n+1}{n}}\left[1-\left(\frac{r}{R}\right)^{\frac{n+1}{n}}\right]=U_0\left[1-\left(\frac{r}{R}\right)^{\frac{n+1}{n}}\right] \qquad [1.151]$$

with

$$U_0=\left(-\frac{K}{2k}\right)^{\frac{1}{n}}\frac{n}{n+1}R^{\frac{n+1}{n}} \qquad [1.152]$$

where U_0 is the velocity at $r=0$, that is, velocity on the axis of the tube. It is an extremum (maximum).

We verify that $u(r)$ is indeed a monotone decreasing function of r.

4.2) Let us calculate the flow rate Q_v

$$Q_v=\int_0^R U_0\left[1-\left(\frac{r}{R}\right)^{\frac{n+1}{n}}\right]2\pi r\,dr=2\pi U_0\int_0^R\left[1-\left(\frac{r}{R}\right)^{\frac{n+1}{n}}\right]rdr \qquad [1.153]$$

In order to calculate the definite integral

$$\int_0^R\left[1-\left(\frac{r}{R}\right)^{\frac{n+1}{n}}\right]rdr \qquad [1.154]$$

we consider $\xi=\frac{r}{R}$. Then, $r\,dr=R^2\,d\xi$, the limits become 0 and 1, and we have:

$$\int_0^R\left[1-\left(\frac{r}{R}\right)^{\frac{n+1}{n}}\right]rdr=R^2\int_0^1\left(1-\xi^{\frac{n+1}{n}}\right)\xi\,d\xi$$

$$=R^2\left(\frac{\xi^2}{2}-\frac{n}{3n+1}\xi^{\frac{3n+1}{n}}\right)\Bigg|_0^1=\frac{n+1}{6n+2}R^2 \qquad [1.155]$$

$$Q_v = 2\pi R^2 U_0 \frac{n+1}{6n+2} \qquad [1.156]$$

The velocity V_q is by definition:

$$Q_v = \pi R^2 V_q \qquad [1.157]$$

Therefore:

$$V_q = \frac{n}{6n+2} U_0 \qquad [1.158]$$

4.3) After several calculations, from the expression of Q_v, we find K:

$$K = 2k\left(\frac{3n+1}{\pi n}\right)^n R^{-(3n+1)} Q_V^n \qquad [1.159]$$

which can also be written as:

$$K = 2k\left(\frac{3n+1}{\pi n}\right)^n \frac{V_q^n}{R^{n+1}} \qquad [1.160]$$

5) By definition, the coefficient of head loss per unit length can be written as:

$$J = \frac{d}{dx}\left(\frac{u^2}{2g} + \frac{p_G}{\rho g}\right) = \frac{1}{\rho g}\frac{dp_G}{dx} = \frac{K}{\rho g} \qquad [1.161]$$

since u does not vary with x. By replacing K and U_0 with their values, we get:

$$J = \frac{2k}{\rho g}\left(\frac{3n+1}{n}\right)^n R^{-(n+1)} V_q^n \qquad [1.162]$$

$$J = \left[\frac{8k}{\rho}\left(\frac{3n+1}{n}\right)^n 2^n D^{-(n)} V_q^n\right] \frac{V_q^2}{2g(2R)} \qquad [1.163]$$

$$\psi = \frac{2^{n+3}\,k}{\rho}\left(\frac{3n+1}{n}\right)^{n} D^{-(n)}\, V_q^{n-2} = 2^{n+3}\,k\left(\frac{3n+1}{4n}\right)^{n} \frac{k}{\rho\, D^{n}\, V_q^{2-n}} \qquad [1.164]$$

By introducing the equivalent Reynolds number R_{en}:

$$R_{en} = 8^{1-n}\left(\frac{4n}{3n+1}\right)^{n} \frac{\rho V_q^{2-n}\, D^{n}}{k} \qquad [1.165]$$

we can identify:

$$\psi = \frac{64}{R_{en}} \qquad [1.166]$$

Given the adopted definitions, a relation that is formally identical to the one resulting from Poiseuille's law can be found.

If $n=1$, the fluid is Newtonian, k is identical to dynamic viscosity μ and we are within the framework of Poiseuille's theory. We find the classical results:

$$R_{en} = \frac{\rho V_q\, D}{\mu} \qquad [1.167]$$

$$J = \left[\frac{8\mu}{\rho}(4)2DV_q\right]\frac{V_q^2}{2g(2R)} = \psi\frac{V_q^2}{2gD} \qquad [1.168]$$

$$\psi = 2^4\,\mu(1)\frac{\mu}{\rho D V_q} = \frac{64}{R_{en}} \qquad [1.169]$$

EXAMPLE 1.5 (Flow in an annular space).–

We study the laminar flow of an incompressible Newtonian fluid of density ρ and dynamic viscosity μ in the annular space limited by two cylindrical walls of radii R_1 and R_2 $(R_1 < R_2)$. The analysis technique employed here is derived from Poiseuille's theory.

The flow rate Q_V in the annular space is given.

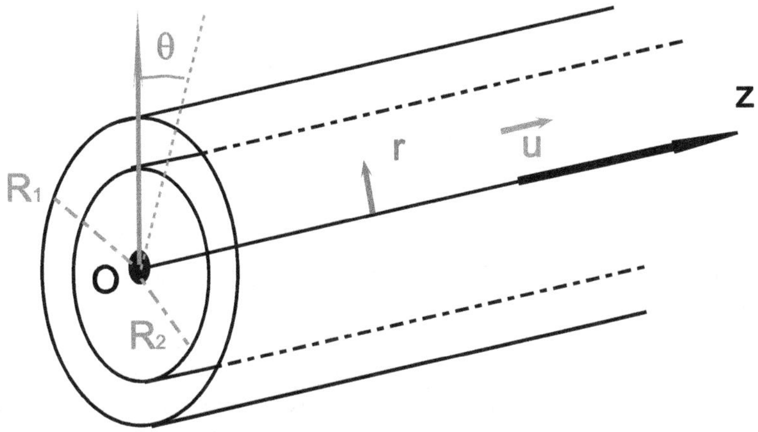

Figure 1.4. *Flow in annular space*

The flow is established: velocity $\vec{V}$ reduces to its axial component u, which depends only on r.

A cylindrical coordinate system (x, r, θ) is used.

To simplify the expressions written, the axis Ox is placed horizontally.

1) Find the differential equation that relates $u(r)$ and the stagnation pressure p_G.

To find it, the law of dynamics is written for a fluid volume limited by two cylinders of radii r and $r+dr$ and two planes perpendicular to Oz (therefore vertical).

2) Writing $u(r)$ as $U_0\, f(r)$, find the function $f(r)$. Find a dimensionless representation of the velocity profile. A constant denoted by K emerges. What does U_0 stand for?

3) Find the expression of $\dfrac{dp_G}{dx}$ as a function of K. How can K be determined? We shall simply indicate the method, without making the calculations.

It is important to note that this method is in fact an implicit application of Euler's theorem (momentum input equals momentum output), and hence an opportunity to understand its significance again.

Solution:

1) This is an established flow. Velocity $\vec{V}$ is reduced to its axial component u.

Let us write the law of dynamics for a fluid volume limited by:

– two cylinders of radii r and $r + dr$,

– two planes perpendicular to Oz (therefore vertical) separated by a distance L

Each particle of fluid of this element moves with constant velocity (vector quantity defined by direction, orientation, hourly speed), therefore its acceleration is null. The acceleration of the fluid volume is null, therefore the resultant of the forces exerted on it is null. Let us project these forces on Ox. Only the surface forces have a non-null projection. These forces consist of viscosity forces and pressure forces. The pressure and gravity forces have the following resultant:

$$-2\pi r\, dr \frac{dp_G}{dx} dx$$

The relation employs p_G, which takes into account the effect of gravity forces on the local pressure. The balance of pressure forces involves opposite surface elements of the same elevation. The result obtained can readily be extended to a tube that is inclined relative to the horizontal. Let us calculate the viscosity forces.

On the inner cylinder of radius r, they are equal to:

$$-2\pi r\, L \frac{du}{dr}$$

On the outer cylinder (radius $r + dr$), they are equal to:

$$2\pi r\, L \frac{du}{dr} + \frac{d}{dr}\left(2\pi r\, L \frac{du}{dr}\right) dr$$

The same sign should be attached to both of these expressions, irrespective of the local orientation of the variation of u with r.

The projection on Ox of the final balance of forces is:

$$-2\pi r\, dr \frac{dp_G}{dx} dx - 2\pi r\, L \frac{du}{dr} + \left[2\pi r\, L \frac{du}{dr} + \frac{d}{dr}\left(2\pi r\, L \frac{du}{dr}\right) dr\right] = 0 \qquad [1.170]$$

After simplification by constant terms and several rearrangements, the sought-for differential equation is:

$$\frac{\mu}{r}\frac{d}{dr}\left(r\frac{du}{dr}\right)=\frac{dp_G}{dx} \qquad [1.171]$$

This system is solved in the classical manner:

The equation equalizes a term exclusively dependent on r with a term exclusively dependent on x. These terms cannot be equal unless they are constant. Let us denote this constant by K. It has been chosen to be positive and it can be predicted that $\frac{dp_G}{dx}$ will be negative.

$$\frac{\mu}{r}\frac{d}{dr}\left(r\frac{du}{dr}\right)=-K \qquad [1.172]$$

$$\frac{dp_G}{dx}=-K \qquad [1.173]$$

2) The equation for u is of second order and it calls for two boundary conditions, which will be the no-slip conditions in R_1 and R_2.

$$r=R_1 \Rightarrow u=0; r=R_2 \Rightarrow u=0 \qquad [1.174]$$

Solving this equation draws on Poiseuille's theory:

$$\frac{d}{dr}\left(r\frac{du}{dr}\right)=-\frac{K}{\mu}r \qquad [1.175]$$

$$r\frac{du}{dr}=-\frac{K}{\mu}\frac{r^2}{2}+C_1 \qquad [1.176]$$

$$\frac{du}{dr}=-\frac{K}{\mu}\frac{r}{2}+\frac{C_1}{r} \qquad [1.177]$$

hence the form of u:

$$u=-\frac{K}{\mu}\frac{r^2}{4}+C_1 Ln r+C_2 \qquad [1.178]$$

The application of boundary conditions yields:

$$-\frac{K}{\mu}\frac{R_1^{\,2}}{4}+C_1\,Ln\,R_1+C_2=0;-\frac{K}{\mu}\frac{R_2^{\,2}}{4}+C_1\,Ln\,R_2+C_2=0 \qquad [1.179]$$

By subtracting the terms of the two equalities, C_1 and C_2 are obtained:

$$C_1\,Ln\frac{R_2}{R_1}-\frac{K}{\mu}\frac{R_2^{\,2}-R_1^{\,2}}{4}=0 \qquad [1.180]$$

$$C_1=\frac{K}{4\mu}\frac{R_2^{\,2}-R_1^{\,2}}{Ln\frac{R_2}{R_1}} \qquad [1.181]$$

$$C_2=\frac{K}{4\mu}R_2^2-\frac{K}{4\mu}\frac{R_2^{\,2}-R_1^{\,2}}{Ln\frac{R_2}{R_1}}Ln\,R_2=\frac{K}{4\mu}R_1^2-\frac{K}{4\mu}\frac{R_2^{\,2}-R_1^{\,2}}{Ln\frac{R_2}{R_1}}Ln\,R_1 \qquad [1.182]$$

$$C_2=\frac{K}{4\mu}\frac{R_2^2+R_1^2}{2}-\frac{K}{4\mu}\frac{R_2^{\,2}-R_1^{\,2}}{Ln\frac{R_2}{R_1}}\frac{Ln\,R_2R_1}{2} \qquad [1.183]$$

Or the alternative form:

$$C_2=\frac{K}{4\mu}\left[R_2^2-\frac{R_2^{\,2}-R_1^{\,2}}{Ln\frac{R_2}{R_1}}Ln\,R_2\right]=\frac{K}{4\mu}\left[R_1^2-\frac{R_2^{\,2}-R_1^{\,2}}{Ln\frac{R_2}{R_1}}Ln\,R_1\right] \qquad [1.184]$$

$$C_2=\frac{K}{4\mu}\left[\frac{R_2^2+R_1^2}{2}-\frac{R_2^{\,2}-R_1^{\,2}}{Ln\frac{R_2}{R_1}}\frac{Ln\,R_2R_1}{2}\right] \qquad [1.185]$$

There are three possible expressions for C_2. The first two use the boundary condition in R_1 or in R_2, while the third is the half-sum of the previous two. It is worth noting that since the flow has no point on the axis of the tube, there is no reason for cancelling out the logarithmic term. This will render further calculations more complicated.

3) In Poiseuille's calculation, which treats the flow in a tube, three constants emerge: C_1, C_2 and K. Two pieces of information are used: a boundary condition (at the wall) which yields C_1, a physical remark on the necessity to have a finite flow rate $(C_2 = 0)$ and an "integral" condition, the flow rate value, which yields K. This solving method is quite unusual and may confuse some readers. It is unexpected to see three constants emerging when solving a second-order differential equation. This is just a paradox, which, as always, is a wrongly posed question. In fact, there are two coupled differential equations, a second order one for u and a first order one for p_G. Therefore, three constants are indeed necessary.

In the present problem, there are also three constants: two integration constants, C_1 and C_2, and a third constant, K. Two boundary conditions have been used. Similarly to the "classical" Poiseuille's theory, K will result from the calculation of the volume flow rate Q_V:

$$Q_V = \int_{R_1}^{R_2} u(r) 2\pi r \, dr = 2\pi \int_{R_1}^{R_2} \left(-\frac{K}{\mu}\frac{r^2}{4} + C_1 \, Lnr + C_2 \right) r \, dr \qquad [1.186]$$

C_1 and C_2 are known, therefore K can be determined.

1.6. Outer boundary layers: integral methods

1.6.1. *Principle of the integral method*

Integral methods, which probably are not very well known, are an elegant approach to solve certain transfer problems (including heat and mass transfer) rapidly in laminar and turbulent boundary layers.

They save efforts in terms of direct solving of the equations when the searched-for results are limited to the scales of the boundary layer and to wall transfer. As will be seen, it is sufficient to suppose a self-similarity of profiles; the accuracy will depend on how close the considered profile is to the real one. This is how profiles that are close to the one resulted from the Blasius analysis can be reintroduced in a heat transfer problem with high success rate.

Similarly, in turbulent regime, the so-called "power law" profiles will prove particularly useful. These methods will not be presented here, as we focus on dynamic boundary layers. It should be kept in mind that these methods are equally effective in heat and mass transfer problems, including in the presence of chemical reactions in the walls. Moreover, the reader might be surprised by the attention given here to laminar layers, the value of which is rather academic than practical, from a technological perspective. If time is deservedly devoted, solving the proposed exercises will challenge the reader to consider lines of reasoning that he/she is not always accustomed to. The study of boundary layer extensively calls for physicist skills in understanding mechanics and practicing orders of magnitude. The latter aspect is decisive when handling the equations, which may prove impracticable in the absence of reasonable simplification.

Integral methods for an outer boundary layer in Cartesian coordinates:

$$\frac{\partial u}{\partial x}+\frac{\partial v}{\partial y}=0 \quad [1.187]$$

$$u\frac{\partial u}{\partial x}+v\frac{\partial u}{\partial y}=-\frac{1}{\rho}\frac{dp_G}{dx}+\nu\frac{\partial^2 u}{\partial y^2}=U_e\frac{dU_e}{dx}+\nu\frac{\partial^2 u}{\partial y^2} \quad [1.188]$$

Here, the possibility of a variation of the outer flow velocity[1] $U_e(x)$ is taken into account.

The identity $-\frac{1}{\rho}\frac{dp_G}{dx}=U_e\frac{dU_e}{dx}$ [1.189]

has been used here. Let us recall (see Chapter 1, section 1.3.1 in [LED 17]) that here we have Bernoulli's equation for outer flow, which is considered perfect. Moreover, we are within the framework of boundary layer approximations; the flow being "very close" to a parallel flow, p_G y is constant in any plane perpendicular to the flow lines. Let us integrate the continuity and impulse equations on y between 0 and infinity:

The first one yields:

$$\int_0^\infty \frac{\partial u}{\partial x}dy=-\int_0^\infty \frac{\partial v}{\partial y}dy=-v_\infty \quad [1.190]$$

1 Integral methods have to a large extent been developed by von Kármán, notably for treating problems of interaction between a real fluid and obstacles. In this case, the potential (outer) flow is eminently variable (see the reminders on this subject in Chapter 3 of [LED 17]).

v_∞ is not null (the flow is "deflected" to the outside, because of the "deficit" of mass flow rate in the boundary layer):

$$\int_0^\infty \left(u\frac{\partial u}{\partial x} + v\frac{\partial u}{\partial y} \right) dy = \int_0^\infty \nu \frac{\partial^2 u}{\partial y^2} dy = \nu \frac{\partial u}{\partial y}\Big|_0^\infty = -\frac{1}{\rho}\mu\frac{\partial u}{\partial y} + U_e\frac{dU_e}{dx} = \frac{\tau_w}{\rho} + U_e\frac{dU_e}{dx} \qquad [1.191]$$

where τ_W is the shear stress. The friction coefficient C_f is defined as:

$$C_f = \frac{2\tau_w}{\rho U_e^2} \qquad [1.192]$$

$$\int_0^\infty \left(u\frac{\partial u}{\partial x} + v\frac{\partial u}{\partial y} \right) dy = \nu \frac{\partial u}{\partial y}\Big|_0^\infty = -\frac{1}{\rho}\mu\frac{\partial u}{\partial y} = \frac{\tau_w}{\rho} \qquad [1.193]$$

$$\int_0^\infty \left(u\frac{\partial u}{\partial x} + v\frac{\partial u}{\partial y} \right) dy = \int_0^\infty \left(\frac{\partial u^2}{\partial x} - u\frac{\partial u}{\partial x} + \frac{\partial uv}{\partial y} - u\frac{\partial v}{\partial y} \right) dy = \int_0^\infty \left(\frac{\partial u^2}{\partial x} + \frac{\partial uv}{\partial y} \right) dy \qquad [1.194]$$

$$\text{As } -u\frac{\partial u}{\partial x} - u\frac{\partial v}{\partial y} = -u\left(\frac{\partial u}{\partial x} + \frac{\partial v}{\partial y} \right) = 0 \qquad [1.195]$$

given the continuity equation. Moreover:

$$\int_0^\infty \frac{\partial uv}{\partial y} dy = U_e v_\infty = -U_e \int_0^\infty \frac{\partial u}{\partial x} dy \qquad [1.196]$$

$$\int_0^\infty \left(u\frac{\partial u}{\partial x} + v\frac{\partial u}{\partial y} \right) dy = = \int_0^\infty \left(\frac{\partial u^2}{\partial x} + \frac{\partial uv}{\partial y} \right) dy = \int_0^\infty \left(\frac{\partial u^2}{\partial x} - U_e\frac{\partial u}{\partial x} \right) dy = \frac{\tau_w}{\rho} \qquad [1.197]$$

$$\int_0^\infty \left(\frac{\partial u^2}{\partial x} - U_e\frac{\partial u}{\partial x} \right) dy = \frac{d}{dx}\int_0^\infty (u^2 - U_e u)\, dy = \frac{\tau_w}{\rho} \qquad [1.198]$$

These expressions are dimensionless when divided by U_e^2. Momentum thickness is defined as:

$$\delta_2 = \int_0^\infty \left(1 - \frac{u}{U_e} \right) \frac{u}{U_e} dy \qquad [1.199]$$

and we introduce the friction coefficient $C_f = \dfrac{2\tau_W}{\rho U_e^2}$ [1.200]

This finally leads to an expression given by von Kármán:

$$\frac{d\delta_2}{dx} = \frac{C_f}{2} \qquad [1.201]$$

The integral method is deduced from this expression: given a hypothesis of form for the profile $\dfrac{u}{U_e} = f\left(\dfrac{y}{\delta(x)}\right)$ [1.202]

a more or less approximate value is calculated, depending on the validity of the expression chosen for δ_2 as a function of $\delta(x)$, which, through derivation, leads to calculate the friction coefficient. Let us note that this expression imposes a flow self-similarity. The effectiveness of the method results from the integration over the boundary layer, which minimizes the effects of an error on the profile, compared to a direct calculation of the wall stress at the wall by local derivation at $y = 0$. In turbulent regime, the same method will be applied to the mean velocity $\overline{u}(x, y)$.

1.6.2. *Applications of integral methods*

EXAMPLE 1.6 (Laminar boundary layer of a Newtonian fluid).–

Let us consider the boundary layer developed by a flat plate in the path of wind in a uniform flow of non-Newtonian fluid. The velocity U_e of the outer flow is constant.

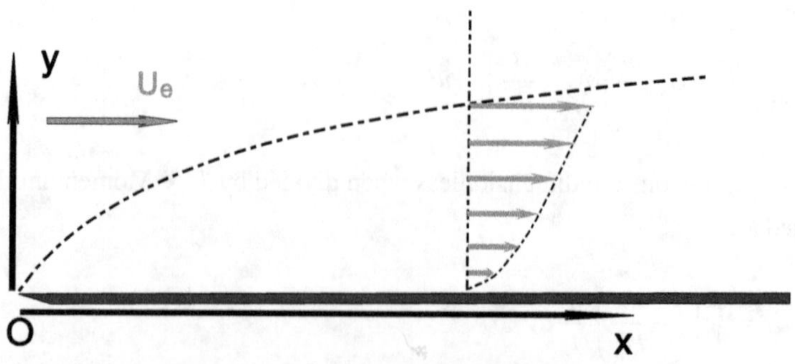

Figure 1.5. *Flat plate in the path of wind*

We shall evaluate the evolution of the thickness of the boundary layer $\delta(x)$ and the friction (through the value of the friction coefficient C_f) using the integral method. Three approximations of the form of velocity profile are proposed. By applying the integral method for the three forms of profile hypothesized below, give in each instance an approximate expression for the evolution of the thickness of the boundary layer, using the ratio $\dfrac{\delta(x)}{x}$ and the friction coefficient $C_f(x)$. Compare these results with those of the Blasius analysis. Which one of these calculations do you find more satisfactory?

1) The first approximation is the most elementary we could imagine

$$\text{for } y \leq \delta; \frac{u}{U_e} = \frac{y}{\delta} \qquad [1.203.a]$$

$$\text{for } y > \delta; \frac{u}{U_e} = 1 \qquad [1.203.b]$$

2) The second approximation relies on a more realistic profile of the boundary layer

$$\text{for } y \leq \delta; \frac{u}{U_e} = \frac{3y}{2\delta} - \frac{1}{2}\left(\frac{y}{\delta}\right)^3 \qquad [1.204.a]$$

$$\text{for } y > \delta; \frac{u}{U_e} = 1 \qquad [1.204.b]$$

3) The third approximation, which will prove equally realistic, takes the following form:

$$\text{for } y \leq \delta; \frac{u}{U_e} = \sin\left(\frac{\pi}{2}\frac{y}{\delta}\right) \qquad [1.205.a]$$

$$\text{for } y > \delta; \frac{u}{U_e} = 1 \qquad [1.205.b]$$

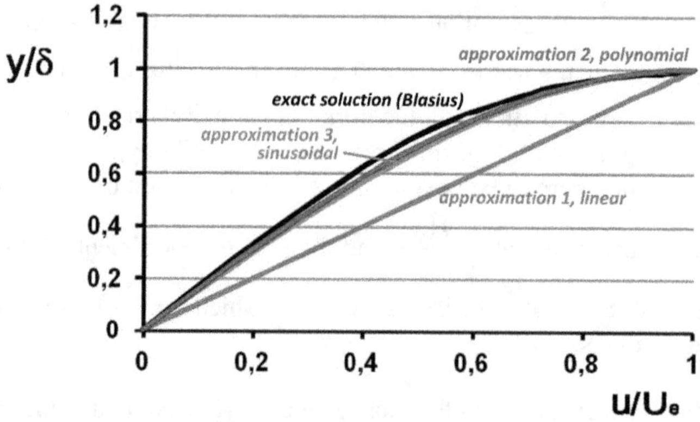

Figure 1.6. *Various approximations of the Blasius profile for boundary layer. For a color version of this figure, see www.iste.co.uk/ledoux/fluid.zip*

Solution:

Let us recall von Kármán equation, which expresses the fundamental law of dynamics in an integral form. In the case of a flow without outer velocity gradient, this equation can be written as:

$$\frac{d\delta_2}{dx}=\frac{C_f}{2} \quad [1.206]$$

where

$$\delta_2 = \int_0^\infty \left(1-\frac{u}{U_e}\right)\frac{u}{U_e}dy \quad [1.207]$$

$$C_f = \frac{2\tau_w}{\rho U_e^2} \quad [1.208]$$

1) Linear approximation

$$\text{for } y \le \delta;\frac{u}{U_e}=\frac{y}{\delta} \quad [1.209.a]$$

$$\text{for } y > \delta;\frac{u}{U_e}=1 \quad [1.209.b]$$

The definition of momentum thickness δ_2 leads to finding its relation with the boundary layer thickness δ:

$$\delta_2 = \int_0^\infty \left(1-\frac{u}{U_e}\right)\frac{u}{U_e}dy = \int_0^\delta \left(1-\frac{y}{\delta}\right)\frac{y}{\delta}\delta\, d\frac{y}{\delta} \qquad [1.210]$$

Let us make the change of variables $\xi = \frac{y}{\delta}$. We integrate from 0 to 1, because the integrand is null beyond $y = \delta$:

$$\delta_2 = \delta \int_0^1 (1-\xi)\xi\, d\xi \qquad [1.211]$$

$$\delta_2 = \delta\left(\frac{\xi^2}{2}-\frac{\xi^3}{3}\right)\Bigg|_0^1 = \frac{\delta}{6} \qquad [1.212]$$

Moreover, the slope of the velocity profile allows us to write the wall shear stress, and then C_f.

NOTE.– It is at this level that the main source of error in this type of calculation is found.

$$\tau_W = \mu \frac{\partial u}{\partial y}\bigg|_{y=0} = \mu\frac{\partial}{\partial y}U_e\frac{u}{U_e}\bigg|_{y=0} = \frac{\mu U_e}{\delta} \qquad [1.213]$$

It is at this level that the main source of error in this type of calculation is found. The derivation depends in effect on the chosen form of profile. It is particularly worth noting that the approximations 2 and 3 are far more adapted from this point of view. In this case, we note that the shear stress is in fact constant in the boundary layer:

$$C_f = \frac{2\tau_w}{\rho U_e^2} \qquad [1.214]$$

$$C_f = \frac{2\mu U_e}{\rho U_e^2\, \delta} = \frac{2\mu}{\rho U_e \delta} \qquad [1.215]$$

Von Kármán equation can then be written as:

$$\frac{1}{6}\frac{d\delta}{dx}=\frac{\mu}{\rho U_e \delta} \qquad [1.216]$$

$$\delta\frac{d\delta}{dx}=\frac{6\mu}{\rho U_e} \qquad [1.217]$$

By introducing the axial Reynolds number $R_x=\frac{U_e x}{\nu}=\frac{\mu U_e x}{\rho}$, and taking into account a boundary layer whose thickness is null at the origin, the differential equation readily yields:

$$\frac{\delta^2}{2}=\frac{6\mu x}{\rho U_e} \qquad [1.218]$$

$$\delta(x)=\sqrt{\frac{12\nu x}{U_e}} \qquad [1.219]$$

$$\frac{\delta(x)}{x}=\frac{3,46}{R_x^{1/2}} \qquad [1.220]$$

C_f can then be written as:

$$C_f=\frac{2\mu}{\rho U_e \delta}=\frac{2\nu}{U_e}\sqrt{\frac{U_e}{12\nu x}} \qquad [1.221]$$

$$\frac{C_f}{2}=\frac{\nu}{U_e}\sqrt{\frac{U_e}{12\nu x}}=\frac{0,289}{R_x^{1/2}} \qquad [1.222]$$

These results are comparable to those of Blasius:

$$\frac{\delta(x)}{x}=\frac{4,95}{R_x^{1/2}};\frac{C_f}{2}=\frac{0,332}{R_x^{1/2}} \qquad [1.223]$$

We shall note that the form of $\frac{\delta(x)}{x}$ and $\frac{C_f}{2}$, up to the constants, is found.

This comparison yields an error of 30% on δ and 13% on the friction coefficient C_f. We shall note that this relatively rough approximation yields a result that is already acceptable in the case of friction, which is the most important from a practical standpoint.

2) Polynomial approximation

$$\text{for } y \le \delta;\ \frac{u}{U_e} = \frac{3y}{2\delta} - \frac{1}{2}\left(\frac{y}{\delta}\right)^3 \quad [1.224.a]$$

$$\text{for } y > \delta;\ \frac{u}{U_e} = 1 \quad [1.224.b]$$

The same reasoning steps as above are applicable

Let us establish the relation between momentum thickness δ_2 and boundary layer thickness δ.

$$\delta_2 = \int_0^\infty \left(1 - \frac{u}{U_e}\right)\frac{u}{U_e}dy = \int_0^\delta \left(1 - \frac{3y}{2\delta} + \frac{1}{2}\left(\frac{y}{\delta}\right)^3\right)\left[\frac{3y}{2\delta} - \frac{1}{2}\left(\frac{y}{\delta}\right)^3\right]\delta d\frac{y}{\delta} \quad [1.225]$$

Let us make the change of variables $\xi = \frac{y}{\delta}$. We integrate from 0 to 1, because the integrand is null beyond $y = \delta$:

$$\delta_2 = \delta \int_0^1 \left(1 - \frac{3\xi}{2} + \frac{1}{2}\xi^3\right)\left[\frac{3\xi}{2} - \frac{1}{2}\xi^3\right] d\xi \quad [1.226]$$

By expanding the product of brackets in the integrand and making a term-by-term integration of the monomials obtained, we have:

$$\delta_2 = \delta \left[-\frac{\xi^7}{28} + \frac{3\xi^5}{10} - \frac{\xi^4}{8} - \frac{9\xi^3}{12} + \frac{3\xi^2}{4}\right]_0^1 \quad [1.227]$$

$$\delta_2 = 0.139\,\delta \quad [1.228]$$

Moreover, the slope of the velocity profile allows us to write the wall shear stress, then C_f:

$$\tau_W = \mu \frac{\partial u}{\partial y}\bigg|_{y=0} = \mu \frac{\partial}{\partial y} U_e \frac{u}{U_e}\bigg|_{y=0}$$

$$= \mu \frac{\partial}{\partial y} U_e \left(\frac{3y}{2\delta} - \frac{1}{2}\left(\frac{y}{\delta}\right)^3 \right)\bigg|_{y=0} = \mu U_e \left(\frac{3y}{2\delta} - \frac{3}{2}\frac{y^2}{\delta^3} \right)\bigg|_0^1 \quad [1.229]$$

$$\tau_W = \frac{3\mu U_e}{2\delta} \quad [1.230]$$

$$\frac{C_f}{2} = \frac{\tau_w}{\rho U_e^2} = \frac{3\nu}{2\delta U_e} \quad [1.231]$$

Then, von Kármán equation can be written as:

$$0.139 \frac{d\delta}{dx} = \frac{3\nu}{2\delta U_e} \quad [1.232]$$

$$\frac{d\delta^2}{dx} = \frac{21.58\nu}{U_e}; x = 0; \delta = 0 \quad [1.233]$$

By introducing the axial Reynolds number $R_x = \frac{U_e x}{\nu} = \frac{\mu U_e x}{\rho}$, the differential equation readily yields:

$$\delta^2 = 21{,}58 \frac{\nu x}{U_e} \quad [1.234]$$

$$\delta(x) = \sqrt{\frac{21{,}58\, \nu x}{U_e}} = 4{,}64 \sqrt{\frac{\nu x}{U_e}} \quad [1.235.a]$$

$$\frac{\delta(x)}{x} = \frac{4{,}64}{R_x^{1/2}} \quad [1.235.b]$$

Then, the friction coefficient comes down to

$$\frac{C_f}{2} = = \frac{3\nu}{2\delta U_e} = \frac{3\nu}{2*4,64\ U_e}\sqrt{\frac{U_e}{\nu x}} \qquad [1.236.a]$$

$$\frac{C_f}{2} = = \frac{0,322}{R_x^{1/2}} \qquad [1.236.b]$$

This comparison yields an error of 6,3% on δ and 2,9% on the friction coefficient C_f. It can be noted that the form of profile adopted, which is close to the Blasius profile, gives an excellent approximation.

3) Sinusoidal approximation

$$\text{For } y \leq \delta; \frac{u}{U_e} = \sin\left(\frac{\pi y}{2\delta}\right) \qquad [1.237.a]$$

$$\text{For } y > \delta; \frac{u}{U_e} = 1 \qquad [1.237.b]$$

The same approach as previously discussed is applicable.

Let us explore the relation between momentum thickness δ_2 and border layer thickness δ:

$$\delta_2 = \int_0^\infty \left(1 - \frac{u}{U_e}\right)\frac{u}{U_e} dy = \int_0^\delta \left(1 - \sin\left(\frac{\pi}{2}\frac{y}{\delta}\right)\right)\sin\left(\frac{\pi}{2}\frac{y}{\delta}\right) dy \qquad [1.238]$$

We consider $\xi = \dfrac{y}{\delta}$ [1.239]

$$\delta_2 = \delta \int_0^1 \left(1 - \sin\left(\frac{\pi}{2}\xi\right)\right)\sin\left(\frac{\pi}{2}\xi\right) d\xi \qquad [1.240]$$

The integration is made from 0 to 1, because the integrand is null beyond $y = \delta$.

Moreover, let us recall that $\cos 2\alpha = \cos^2\alpha - \sin^2\alpha = 2\cos^2\alpha - 1$

$$\delta_2 = \delta \int_0^1 \left(\sin\left(\frac{\pi}{2}\xi\right) - \sin^2\left(\frac{\pi}{2}\xi\right) \right) d\xi$$

$$= \delta \int_0^1 \left(\sin\left(\frac{\pi}{2}\xi\right) - \frac{1-\cos 2\frac{\pi}{2}\xi}{2} \right) d\xi \qquad [1.241]$$

$$\delta_2 = \delta \left[-\frac{2}{\pi}\cos\left(\frac{\pi}{2}\xi\right)\Big|_0^1 - \frac{\xi}{2}\Big|_0^1 + \frac{\sin\pi\xi}{2\pi}\Big|_0^1 \right]$$

$$= \delta\left(\frac{2}{\pi} - \frac{1}{2}\right) + 0 = \frac{4-\pi}{2\pi}\delta = 0.137\delta \qquad [1.242]$$

$$\delta_2 = 0{,}137\delta \qquad [1.243]$$

The friction is related to the thickness of the boundary layer by:

$$\tau_W = \mu \frac{\partial u}{\partial y}\Big|_{y=0} = \mu \frac{\partial}{\partial y} U_e \frac{\pi}{2}\cos\left(\frac{\pi}{2}\xi\right)\Big|_{y=0} = \frac{\pi}{2}\frac{\mu U_e}{\delta} \qquad [1.244]$$

$$C_f = \frac{2\tau_w}{\rho U_e^2} = \pi \frac{\nu}{U_e \delta} \qquad [1.245]$$

$$\frac{C_f}{2} == \frac{\pi}{2}\frac{\nu}{U_e \delta} \qquad [1.246]$$

von Kármán equation can then be written as:

$$0.137\frac{d\delta}{dx} = \frac{\pi}{2}\frac{\nu}{U_e \delta} \qquad [1.247]$$

$$\delta\frac{d\delta}{dx} = 11.47\frac{\nu}{U_e \delta} \qquad [1.248]$$

$$\frac{d\delta^2}{dx} = 22.93\frac{\nu}{U_e \delta};\ x = 0;\ \delta = 0 \qquad [1.249]$$

This simple differential equation associated to its boundary condition yields:

$$\delta^2 = 22,93\frac{\nu x}{U_e} \quad [1.250]$$

$$\delta = 4,79\sqrt{\frac{\nu x}{U_e}} \quad [1.251]$$

or, still further, by introducing the axial Reynolds number $R_x = \frac{U_e x}{\nu}$,

$$\frac{\delta}{x} = 4,79\sqrt{\frac{\nu x}{U_e}} = \frac{4,79}{\sqrt{R_x}} \quad [1.252]$$

The friction coefficient C_f then results as:

$$\frac{C_f}{2} = \frac{\pi}{2}\frac{\nu}{U_e \delta} = 0,328\sqrt{\frac{\nu}{U_e x}} = \frac{0,328}{\sqrt{R_x}} \quad [1.253]$$

These results are comparable to those of Blasius. The expressions of $\frac{\delta}{x}$ and C_f are quite identical in the form to those of Blasius. By adopting this third profile, the thickness of the boundary layer is underestimated by $3,2\%$ and the friction coefficient by $1,2\%$. This validates another very good approximation, which introduces circular functions in the calculation.

EXAMPLE 1.7 (Plate in the path of wind. Friction with injection of wall fluid).–

The objective is to study the boundary layer produced by a far-off uniform flow of velocity U_e and a flat porous plate placed in the path of wind, parallel to the streamlines of potential flow. The fluid is incompressible with density ρ and dynamic viscosity μ. All these physical properties are constant. Fluid is taken out all along the plate with a constant velocity of module W. The flow being plane, a system of axes Oxy (Ox along the plate) will be used as frame of reference. The origin O is chosen on the leading edge. The velocity has two components u and v, directed along Ox and Oy, respectively.

Starting from a certain abscissa x_0, the flow is supposed to be self-similar, meaning that the longitudinal component of the local velocity depends only on the distance to the wall, y: $u = u(y)$. The region near the leading edge ($x < x_0$) will be called non-similar region. The problem will be solved first by an analytical method, and then by an integral method.

1) Analytical method

Only the similar region of the flow is of interest here.

1.1) Deduce from the expression of continuity the profile of the lateral component v of the velocity in the boundary layer.

1.2) Deduce from it the simple form taken by the impulse equation.

1.3) Find the profiles of velocity and friction.

a) Find the profile of velocity $u(y)$

b) Deduce from it the thickness of the dynamic boundary layer.

c) Write the expressions of the wall shear stress τ_W and the friction coefficient C_f.

2) Integral method

2.1) Unless it has not already been solved, treat question 1.1

2.2) Establish for this problem the relation between δ_2, C_f and W. This relation will be established for the whole boundary layer, including the non-similar region. Let us recall that in this region $u = u(x,y)$; $v = v(x,y)$. The integral method will be used in order to find the friction coefficient C_f in the self-similar region. For this purpose, the previously described analysis used for establishing von Kármán relation will be resumed, taking into account the fact that v is not null at the wall. Moreover, the following considerations are made:

The thickness of the dynamic boundary layer δ is constant in this region.

The profile of $u(y)$ is self-similar:

$$\frac{u}{U_e} = f\left(\frac{y}{\delta}\right) \qquad [1.254]$$

where $f\left(\frac{y}{\delta}\right)$ is a function that needs not be *a priori* given.

Compare the expression found for $C_f(x)$ to the one obtained at question 1.

Solution:

1) Analytical method

1.1) Profile of $v(y)$

For a two-dimensional plane and incompressible flow, the continuity equation is written as:

$$\frac{\partial u}{\partial x}+\frac{\partial v}{\partial y}=0 \qquad [1.255]$$

Here, it comes down to:

$$u=u(y);\frac{\partial u}{\partial x}=0 \qquad [1.256]$$

$$\frac{\partial v}{\partial y}=0 \qquad [1.257]$$

v is constant in the boundary layer, therefore y will have a fixed value at the wall:

$$v=-W \qquad [1.258]$$

1.2) The impulse equation for the two-dimensional plane and incompressible boundary layer is written in a simplified form as:

$$u\frac{\partial u}{\partial x}+v\frac{\partial u}{\partial y}=-\frac{1}{\rho}\frac{\partial p}{\partial x}+\nu\frac{\partial^2 u}{\partial y^2} \qquad [1.259]$$

In the absence of a longitudinal gradient of the outer velocity:

$$-\frac{1}{\rho}\frac{\partial p}{\partial x}=U_e\frac{dU_e}{dx}=0 \qquad [1.260]$$

The final form thus results:

$$0-W\frac{\partial u}{\partial y}=\nu\frac{\partial^2 u}{\partial y^2} \quad [1.261]$$

$$y=0;u=0 \quad [1.262]$$

1.3) Finding velocity and friction

a) Velocity profile

The equation in $u(y)$ is readily integrated:

$$0-W\frac{\partial u}{\partial y}=\nu\frac{\partial^2 u}{\partial y^2} \quad [1.263]$$

$$\frac{\dfrac{\partial^2 u}{\partial y^2}}{\dfrac{\partial u}{\partial y}}=-\frac{W}{\nu} \quad [1.264]$$

$$Ln\left(\frac{\partial u}{\partial y}\right)=-\frac{W}{\nu}y+LnC_1 \quad [1.265]$$

Taking the exponential of the two terms:

$$\frac{\partial u}{\partial y}=C_1\exp-\frac{W}{\nu}y \quad [1.266]$$

$$u=\int_0^y C_1\exp-\frac{W}{\nu}dy+C_2 \quad [1.267]$$

$$u=-\frac{\nu}{W}C_1\ \exp-\frac{W}{\nu}y\Big|_0^y=-\frac{\nu}{W}C_1\left(\ \exp-\frac{W}{\nu}y-1\Big|\right)+C_2 \quad [1.268]$$

Let us note that we could have also written $u=\int_{C_2}^y C_1\exp-\frac{W}{\nu}dy$, which obviously leads to the same result. Applying the boundary conditions:

$$y=0;\ u=0;C_2=0 \quad [1.269]$$

$$y \to \infty; u \to U_e;\ U_e = -\frac{\nu}{W} C_1 \left(0 - 1\right) + C_2 = \frac{\nu}{W} C_1;\ C_1 = \frac{W U_e}{\nu} \qquad [1.270]$$

Finally:

$$u = U_e \left(1 - \exp - \frac{W}{\nu} y\right) \qquad [1.271]$$

Let us note that $\frac{W}{\nu} y$ is a Reynolds number.

b) Thickness of the boundary layer

According to the definition of the limit of the boundary layer:

$$y = \delta; \frac{u}{U_e} = 0{,}99 \qquad [1.272]$$

$$1 - \exp - \frac{W}{\nu} \delta = 0{,}99 \qquad [1.273]$$

$$\delta = 4{,}605 \frac{\nu}{W} \qquad [1.274]$$

Or if the result is expressed in dimensionless form, by means of a Reynolds number:

$$\frac{W\delta}{\nu} = 4{,}605 \qquad [1.275]$$

We note that in the self-similar region of the flow, the thickness of the boundary layer is constant, in the same way as the Reynolds number $\frac{W\delta}{\nu}$

c) Friction

The wall shear stress can be calculated directly from the velocity profile:

$$\tau = \mu \frac{du}{dy} = \mu \frac{d}{dy} U_e \left(1 - \exp - \frac{W}{\nu} y\right)_{y=0} = \rho W U_e \qquad [1.276]$$

$$C_f = \frac{2\tau_W}{\rho U_e^2} = 2\frac{W}{U_e} \qquad [1.277]$$

2) Integral method

The integral method was presented in 1.6.1 and in the previous examples for a fluid-proof plate. The analysis will be slightly more complicated for the present case. The Von Kármán expression is not directly applicable. This analysis must be resumed.

2.1) Let us recall that in 1.1, it was established that the lateral component of velocity is constant in the boundary layer:

$$v = -W \qquad [1.278]$$

2.2) Integral method for an outer boundary layer in Cartesian coordinates. Let us apply the reasoning of 1.6.1 in the case of the plate with transpiration:

$$\frac{\partial u}{\partial x} + \frac{\partial v}{\partial y} = 0 \qquad [1.279]$$

$$u\frac{\partial u}{\partial x} + v\frac{\partial u}{\partial y} = \nu\frac{\partial^2 u}{\partial y^2} \qquad [1.280]$$

Let us integrate the continuity and impulse equations over y between 0 and infinity:

The first one yields:

$$\int_0^\infty \frac{\partial u}{\partial x} dy = -\int_0^\infty \frac{\partial v}{\partial y} dy = -\left[v_\infty - v(y=0)\right] = W - v_\infty \qquad [1.281]$$

v_∞ is not null (outer flow deflection, due to the deficit of mass rate flow in the limit boundary):

$$\int_0^\infty \left(u\frac{\partial u}{\partial x} + v\frac{\partial u}{\partial y}\right) dy = \nu \frac{\partial u}{\partial y}\Big|_0^\infty = -\frac{1}{\rho}\mu\frac{\partial u}{\partial y} = \frac{-\tau_w}{\rho} \qquad [1.282]$$

where τ_w is the wall stress.

The friction coefficient C_f is defined as:

$$C_f = \frac{2\tau_w}{\rho U_e^2} \quad [1.283]$$

$$\int_0^\infty \left(u\frac{\partial u}{\partial x} + v\frac{\partial u}{\partial y} \right) dy = \nu \frac{\partial u}{\partial y}\Big|_0^\infty = -\frac{1}{\rho}\mu\frac{\partial u}{\partial y} = \frac{-\tau_w}{\rho} \quad [1.284]$$

$$\int_0^\infty \left(u\frac{\partial u}{\partial x} + v\frac{\partial u}{\partial y} \right) dy = \int_0^\infty \left(\frac{\partial u^2}{\partial x} - u\frac{\partial u}{\partial x} + \frac{\partial uv}{\partial y} - u\frac{\partial v}{\partial y} \right) dy = \int_0^\infty \left(\frac{\partial u^2}{\partial x} + \frac{\partial uv}{\partial y} \right) dy \quad [1.285]$$

since $-u\frac{\partial u}{\partial x} - u\frac{\partial v}{\partial y} = -u\left(\frac{\partial u}{\partial x} + \frac{\partial v}{\partial y} \right)$ is null because of the continuity equation. Moreover:

$$\int_0^\infty \frac{\partial uv}{\partial y} dy = U_e v_\infty = U_e \left(W - \int_0^\infty \frac{\partial u}{\partial x} dy \right) \quad [1.286]$$

$$\int_0^\infty \left(u\frac{\partial u}{\partial x} + v\frac{\partial u}{\partial y} \right) dy = \int_0^\infty \left(\frac{\partial u^2}{\partial x} + \frac{\partial uv}{\partial y} \right) dy = \int_0^\infty \left(\frac{\partial u^2}{\partial x} - U_e\frac{\partial u}{\partial x} \right) dy + W U_e = \frac{\tau_w}{\rho} \quad [1.287]$$

$$\int_0^\infty \left(\frac{\partial u^2}{\partial x} - U_e\frac{\partial u}{\partial x} \right) dy = \frac{d}{dx}\int_0^\infty (u^2 - U_e u)\, dy = \frac{\tau_w}{\rho} - W U_e \quad [1.288]$$

These expressions are rendered dimensionless by dividing the two terms of the equation by U_e^2. The momentum thickness is then defined as:

$$\delta_2 = \int_0^\infty \left(1 - \frac{u}{U_e} \right) \frac{u}{U_e} dy \quad [1.289]$$

and the friction coefficient is introduced:

$$C_f = \frac{2\tau_W}{\rho U_e^2} \quad [1.290]$$

The form of von Kármán equation specific to our problem is finally obtained:

$$\frac{d\delta_2}{dx} = \frac{C_f}{2} - \frac{W}{U_e} \qquad [1.291]$$

Let us establish the relation between δ and δ_2 by using, as previously, its definition:

$$\delta_2 = \int_0^{\infty}\left(1-\frac{u}{U_e}\right)\frac{u}{U_e}dy = \delta\int_0^{\infty}\left(1-f\left(\frac{y}{\delta}\right)\right)f\left(\frac{y}{\delta}\right)d\frac{y}{\delta} = F\delta \qquad [1.292]$$

$$F = \int_0^{\infty}\left(1-f\left(\frac{y}{\delta}\right)\right)f\left(\frac{y}{\delta}\right)d\frac{y}{\delta} \qquad [1.293]$$

F is a definite integral, the value of which, as we shall see, does not need to be known. In effect, *taking into account the invariability of* δ *in the self-similar region*, von Kármán law can be written as:

$$\frac{d\delta_2}{dx} = F\frac{d\delta}{dx} = 0 = \frac{C_f}{2} - \frac{W}{U_e} \qquad [1.294]$$

This obviously yields:

$$C_f = \frac{2W}{U_e} \qquad [1.295]$$

We obtain the result that has already been established by the analytical approach.

1.7. Channels and films

In a monophasic fluid, the outer boundary layers lead to distinguish a boundary layer region, in which the viscosity phenomena are predominant, and an outer flow region, which is often treated as a perfect fluid. In a film or channel, a liquid flow is in contact, through one or more of its boundaries, with a solid wall, and through another boundary it is in contact with a gaseous fluid. At the level of the solid, viscosity forces need to be taken into account. At the level of gas, the gas-to-liquid dynamic viscosity ratios lead to consider that friction is negligible at the limit. Moreover, the thickness of the liquid flow can often be considered constant. Examples of this situation will be found in the following.

EXAMPLE 1.8 (The Pont du Gard in Roman times).–

"Indeed, in my opinion the three most magnificent works of Rome, in which the greatness of her empire is best seen, are the aqueducts, the paved roads and the construction of sewers. I say this with respect not only to the usefulness of the work (concerning which I shall speak in the proper place), but also to the magnitude of the cost...." (Dionysius of Halicarnassus)

The renowned Pont du Gard, near Nîmes, in France, is the vestige of an aqueduct that was built during the 1st Century AD, under the Emperor Claudius, with the purpose of bringing water from the "source d'Eure" spring, near Uzès, to the city of Nîmes, at 50 km distance. The altitude Z of the source is 12.5 m higher than that of Nîmes. Let us conceive a model of this structure.

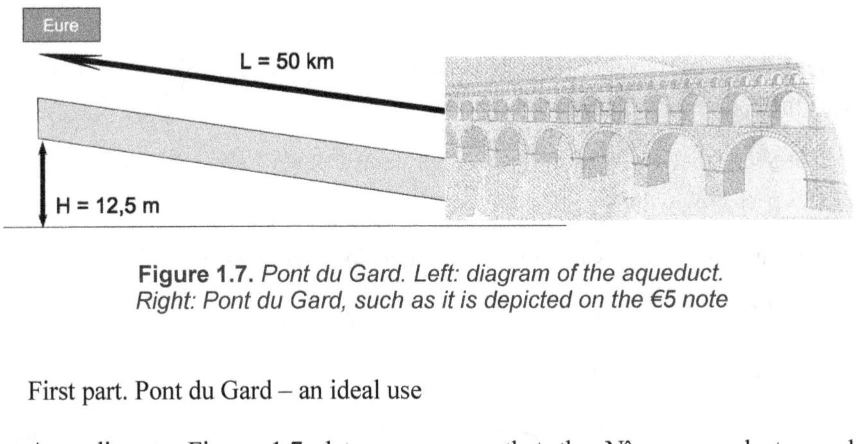

Figure 1.7. *Pont du Gard. Left: diagram of the aqueduct. Right: Pont du Gard, such as it is depicted on the €5 note*

First part. Pont du Gard – an ideal use

According to Figure 1.7, let us suppose that the Nîmes aqueduct can be assimilated to a rectilinear channel of width $l = 1\,m$, between two points whose difference in altitude is $H = 12,5\,m$. The length of the channel is $L = 50\,km$.

The flow in this channel is considered rigorously parallel and of constant depth h. Let viscosity and density of water be $\mu = 1,8.10^{-3}\,Pl$ and $\rho = 1000\,kg.m^{-3}$, respectively. Air will be considered a perfect fluid.

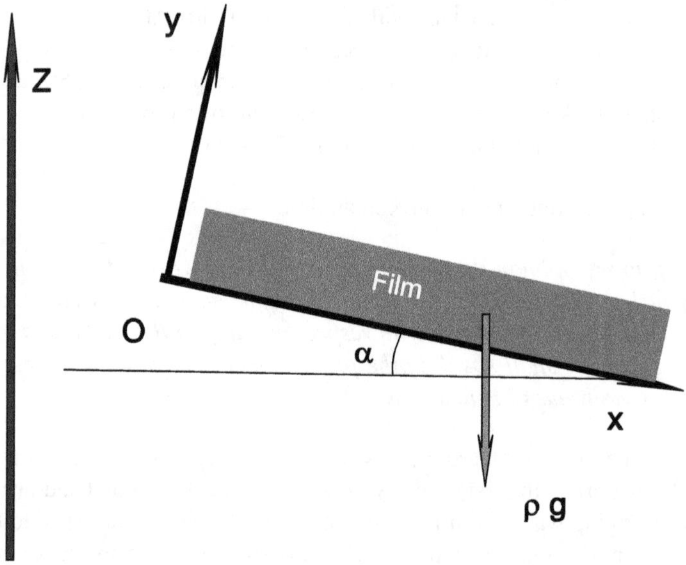

Figure 1.8. *Pont du Gard: the system of coordinates*

1) A frame of reference Ox, Oy is attached to the system, where Ox is located along the bottom of the channel.

1.1) Show that the velocity $\vec{V}(u,v,w)$ comes down to a single component u, which varies only with y.

1.2) Find the form of variation of the stagnation pressure $p_G(x)$.

1.3) Write the differential equation satisfied by $u(y)$.

1.4) Find the expression of $u(y)$.

1.5) Find the expression of the volume rate flow q_v in the channel, in literal form.

1.6) According to ancient texts, the flow rate from the source supplying Nîmes was $36000\, m^3.per\ day$. What was the depth h of the water flow?

Solution:

1) The flow satisfies the continuity equation

1.1) For an incompressible fluid, it should be written in two dimensions:

$$div\vec{V} = 0; \frac{\partial u}{\partial x} + \frac{\partial v}{\partial y} = 0 \quad [1.296]$$

We have considered a parallel film flow. Therefore, velocity reduces to its component parallel to the wall, u, which entails $v = 0$. It follows that

$$\frac{\partial u}{\partial x} + 0 = 0 \quad [1.297]$$

and $u = u(y)$ is a function of only y

1.2) Let us define a vertical axis OZ. This axis should be clearly distinguished from an axis 0z, which is not represented here and which would complete the orthonormal system of reference (Ox, Oy, Oz).

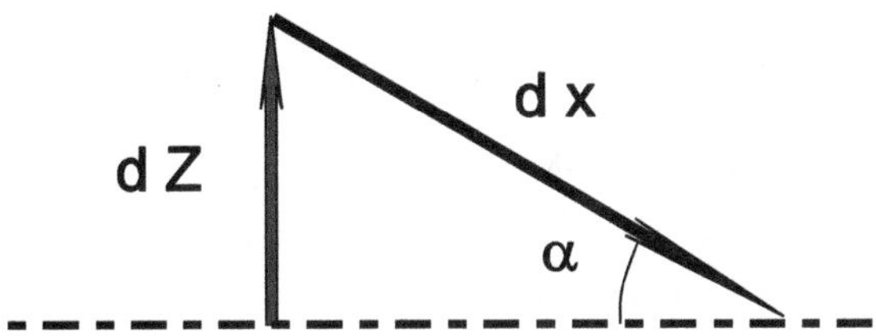

Figure 1.9. *Relation between dz and dx*

Obviously, an angle exists between OZ and Oy. Let us denote by α the angle of the channel relative to the horizontal. It can be readily calculated, knowing that the channel goes down 12.5 m over a length of 50 km:

$$\sin\alpha = \frac{12.5}{50\,000} = 2.5.10^{-4}; \alpha = 1.43.10^{-2}\,° \quad [1.298]$$

Therefore: $dZ = -dx \sin\alpha$ [1.299]

The derivative with respect to x of the stagnation pressure can then be written as:

$$\frac{dp_G}{dx} = \frac{d(p + \rho g Z)}{dx} = \frac{dp}{dx} + \rho g \frac{dZ}{dx} = \frac{dp}{dx} - \rho g \sin\alpha \quad [1.300]$$

1.3) The equation of dynamics, or the impulse equation, for a two-dimensional flow of incompressible Newtonian fluid can be written as:

$$\nu\left[\frac{\partial^2 u}{\partial y^2}\right]+\rho g\sin\alpha=0 \qquad [1.301]$$

u depends only on y, the partial derivatives can be transformed into full derivatives.

$$\nu\left[\frac{d^2 u}{dy^2}\right]+\rho g\sin\alpha=0 \qquad [1.302]$$

The volume forces reduced to gravity and relative to a system of reference whose axis $0x$ is at an angle α with the horizontal can be written as:

$$F_{Vx}=\rho g\sin\alpha \qquad [1.303]$$

Taking into account that both $\frac{\partial u}{\partial x}$ and v are null, the equation of dynamics is finally reduced to two equivalent expressions:

$$-\frac{1}{\rho}\frac{\partial p}{\partial x}+\nu\left[\frac{d^2 u}{dy^2}\right]+\rho g\sin\alpha=0 \qquad [1.304]$$

or:

$$-\frac{1}{\rho}\frac{\partial p_G}{\partial x}+\nu\left[\frac{d^2 u}{dy^2}\right]=0 \qquad [1.305]$$

In any plane perpendicular to the flow lines of the film, the stagnation pressure p_G is constant. Therefore, for any abscissa x, it is equal to its value on the free surface. At this point, the static pressure is equal to the atmospheric pressure p_a. We therefore have:

$$\frac{dp_G}{dx}=\frac{dp}{dx}-\rho g\sin\alpha=\frac{dp_a}{dx}-\rho g\sin\alpha=0-\rho g\sin\alpha \qquad [1.306]$$

The impulse equation can be written as:

$$\mu\left[\frac{d^2 u}{dy^2}\right]+\rho g\sin\alpha=0 \qquad [1.307]$$

The boundary conditions can be written simply as:

a) non-slip condition:

on the bottom of the channel, velocity is null: $y=0; u=0$ [1.308]

b) air is a perfect fluid: the shear stress it applies on the flow surface.

Given the law of action and reaction, the shear stress in water at the free surface is null:

$$y=h;\ \mu\frac{du}{dy}=0 \qquad [1.309]$$

1.4) The second-order linear equation can be readily solved. We proceed by a double integration in y, which will yield the two constants C_1 and C_2 resulted from the boundary conditions:

$$\mu\left[\frac{d^2 u}{dy^2}\right]+\rho g\sin\alpha=0 \qquad [1.310]$$

$$\frac{d^2 u}{dy^2}=-\frac{g}{\nu}\sin\alpha=0 \qquad [1.311]$$

Let us integrate once:

$$\frac{du}{dy}=-\left(\frac{g}{\nu}\sin\alpha\right)y+C_1 \qquad [1.312]$$

Let us integrate a second time:

$$u=-\left(\frac{g}{\nu}\sin\alpha\right)\frac{y^2}{2}+C_1\,y+C_2 \qquad [1.313]$$

Let us now apply the boundary conditions:

$$y = 0; u = 0 = -\left(\frac{g}{\nu}\sin\alpha\right) * \frac{0^2}{2} + C_1 * 0 + C_2; C_2 = 0 \qquad [1.314]$$

$$y = h;\ \mu\frac{du}{dy} = 0; \frac{du}{dy} = 0 = -\left(\frac{g}{\nu}\sin\alpha\right)h + C_1; C_1 = \left(\frac{g}{\nu}\sin\alpha\right)h \qquad [1.315]$$

Given the expression of u :

$$u(y) = -\left(\frac{g}{\nu}\sin\alpha\right)\frac{y^2}{h} + \left(\frac{gh}{\nu}\sin\alpha\right)y \qquad [1.316]$$

This expression can be reformulated under a dimensionless form, by evidencing a reduced velocity $\tilde{u} = \frac{u}{U_0}$ and a reduced ordinate $\xi = \frac{y}{h}$

$$u(y) = -\left(\frac{g}{\nu}\sin\alpha\right)\frac{y^2}{2}\frac{h^2}{h^2} + \left(\frac{gh}{\nu}\sin\alpha\right)y\frac{h}{h} \qquad [1.317]$$

$$u(y) = U_0\left(-\frac{y^2}{2h^2} + \frac{y}{h}\right) = U_0\left(\xi - \frac{\xi^2}{2}\right) \qquad [1.318]$$

$$U_0 = \frac{g\,h^2}{\nu}\sin\alpha \qquad [1.319]$$

$$\tilde{u}(\xi) = \xi - \frac{\xi^2}{2} \qquad [1.320]$$

1.5) To calculate the volume flow rate q_v in the aqueduct, we shall divide a section into small horizontal strips of width l of the aqueduct and height dy, therefore of surface area $dS = l\,dy$, in order to take into account the variation of u with y. It can therefore be written in a classical manner:

$$q_v = \int_0^h u(y)\,dS = \int_0^h u(y)\,l\,dy \qquad [1.321]$$

After a change of variable:

$$\xi = \frac{y}{h};\ d\xi = \frac{dy}{h} \tag{1.322}$$

$$q_v = \int_0^h u(y)\,dS = \int_0^1 u(y)\,lh\,d\xi \tag{1.323}$$

$$q_v = \int_0^1 u(y)\,lh\,d\xi = \int_0^1 U_0\left(\xi - \frac{\xi 2}{2}\right)lh\,d\xi \tag{1.324}$$

$$U_0 = \frac{gh^2}{\nu}\sin\alpha \tag{1.325}$$

$$q_v = lhU_0 \int_0^1 \left(\xi - \frac{\xi 2}{2}\right)d\xi = lhU_0 \left.\frac{\xi 2}{2} - \frac{\xi 3}{6}\right|_0^1 \tag{1.326}$$

The final expression of the flow rate is:

$$q_v = \frac{lhU_0}{3} = \frac{glh^3}{3\nu}\sin\alpha \tag{1.327}$$

1.6) Numerical application

It is known that the flow rate of the source supplying Nîmes was 36000 m^3 per day, or $q_v = \dfrac{36000}{24*3600} = 0,417\,m^3s^{-1}$ [1.328]

Using the numerical values in SI:

$$q_v = \frac{lhU_0}{3} = \frac{glh^3}{3\nu}\sin\alpha \tag{1.329}$$

$$\sin\alpha = 2,5.10^{-4};\ \nu = \frac{1,8.10^{-3}}{1000} = 1.8.10^{-6} \tag{1.330}$$

$$0,417 = \frac{9,81*1*2,5.10^{-4}}{3*1,8.10^{-6}}h^3 \tag{1.331}$$

Therefore, the height of water in the aqueduct results:

$$h = 9,72.10^{-2}\, m = 9,72\, cm \qquad [1.332]$$

Second part. Hackers of the aqueduct

From a purely "academic" perspective, this part of the problem should definitely find its place in Chapter 4. In effect, what follows is a classical flow problem pertaining to the mechanics of perfect fluids (problem related to the one of "the invert"). For the sake of some aestheticism, this example has been related to the previous one, of the film. It is known, particularly from the work "*De Aquaeductu Urbis Romae*" of Frontinus, that despite the formal ban enforced by the Roman emperor and the very harsh repressive measures (among other punishments, those who were caught risked having one hand cutoff), unscrupulous citizens engaged in water diversion.

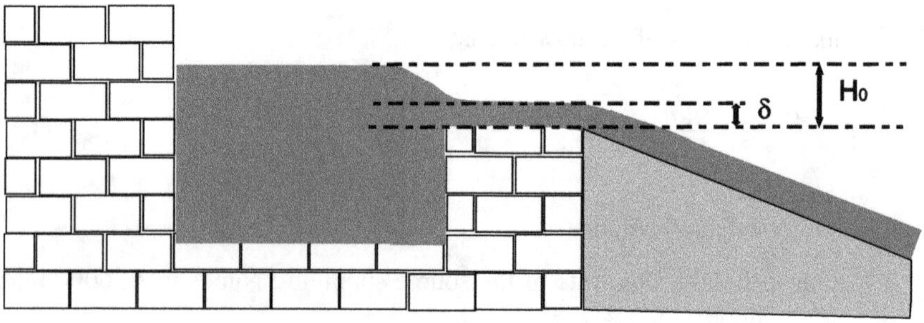

Figure 1.10. *Hacking the aqueduct*

Figure 1.10 represents an instance of water diversion. To estimate the flow rate of the water diversion, we shall study the flow in a particular region where there is a definite flow rate and the fluid can be considered perfect. While legitimate up to an approximation in this case, this hypothesis would obviously not apply to calculating the kinematics of the film. This will not be considered here. It would actually come down to resuming the line of reasoning in part 1 of the example. In this model, the aqueduct is considered a tank whose water height is estimated at $H_0 = 8\,cm$, constant in time (in fact, it is of the order of the height of the film determined in the first part of this example).

NOTE.– We shall verify that water diversion is not done over the whole height of the aqueduct channel.

The water in a tank is considered at rest. Therefore, in this model, the flow in the channel is considered to have no effect on the hydrodynamics of water diversion.

The discharge is considered to lead to the following kinematic configuration: between the "tank" and the "film" flowing into the diversion channel, a rigorously parallel flow is established in a region AB, in which the water depth is δ. At this level, the diversion has a width l_0 (perpendicular to the figure) of 50 cm. The elevations H_0 and δ are illustrated in Figure 1.10.

If δ is not known, it will be subsequently determined. It is simply known that, out of all the values δ arbitrarily considered to take, the only one to be observed is the one leading to the maximum flow rate. In other words, nature automatically "optimizes" the output flow rate.

2.1) Show that the flow is uniform in the AB region. Find the velocity u_f and the volume flow rate q_{vf} of this flow for a hypothetical value of δ.

Then, deduce the observed value of δ.

2.2) Given $H_0 = 8\,cm$, what is the flow rate of the water diversion expressed in m^3 *per day* ?

2.3) It is estimated that during the 3rd Century water diversions had reduced the flow rate of water supply to the city of Nîmes to $13000 m^3$ *per day* ! Making a rough estimation that all water diversions have the same order of magnitude, which we have just calculated, and counting one diversion per family, how many uncivic households were there in the countryside around Nîmes?

Solution:

2.1) Let us attach to the flow in the region AB, which is obviously two-dimensional, a system of axes Ox, Oy. The axis Ox is along the bottom of the channel, and hence it is horizontal in this region. The axis Oy is therefore vertical, and y is an elevation. In this system, H_0 is the elevation of the free surface. In this tank, the fundamental law of hydrostatics is verified:

$$p + \rho g y = Cst = p_a + \rho g H_0 \qquad [1.333]$$

NOTE.– Let us recall that in this "modeling", we forget that the "tank" is in fact a "flow". We shall accept the hypothesis of the tank, which serves the purpose of simplification. On each flow line in the AB flow coming out of the tank, between a point C of the tank and a point D of the flow, the following can be written as:

$$p_C + \rho g y_C + 0 = p_D + \rho g y_D + \rho \frac{u^2(y_D)}{2} \qquad [1.334]$$

The stagnation pressure $p_G = p + \rho g y$ is evidenced, constant in any plane perpendicular to Ox. For any plane perpendicular to Ox, we have:

$$p_G = p + \rho g y = p_a + \rho g \delta \qquad [1.335]$$

Given that the film on AB has a constant thickness δ, the stagnant pressure is constant throughout the flow. A synthesis of what has just been stated gives:

$$p_C + \rho g y_C + 0 = p_a + \rho g H_0 = p_{GD} + \rho \frac{u^2(y_D)}{2} = p_a + \rho g \delta + \rho \frac{u^2(y_D)}{2} \qquad [1.336]$$

$$\rho \frac{u^2(y_D)}{2} = \rho g (H_0 - \delta) \qquad [1.337]$$

$$u(y_D) = \sqrt{2g(H_0 - \delta)} \qquad [1.338]$$

The velocity is therefore constant throughout the AB flow region. This result could have been intuitively predicted by noting that the flow has constant energy. The flow rate $q_{vf}(\delta)$ calculated for the as yet unknown value of δ can be readily deduced as:

$$q_{vf}(\delta) = Su = l_0 \delta \sqrt{2g(H_0 - \delta)} \qquad [1.339]$$

The observed value will be the maximum of this function of δ. For a more convenient calculation, we shall note that this will also be the extremum of the function $\delta^2 (H_0 - \delta)$:

$$\frac{d}{d\delta} \delta^2 (H_0 - \delta) = 2\delta (H_0 - \delta) - \delta^2 = \delta(2H_0 - 3\delta) \qquad [1.340]$$

The derivative is annulled for $\delta = 0$, which corresponds to a null flow rate, hence the extremum is a minimum obtained for $\delta = \frac{2H_0}{3}$ [1.341]

which is the sought-for solution.

2.2) The flow rate of the diversion will consequently be:

$$q_{vf}(\delta) = Su = l_0 \frac{2H_0}{3}\sqrt{2g\left(H_0 - \frac{2H_0}{3}\right)} \quad [1.342]$$

$$q_{vf}(\delta) = l_0 \frac{2H_0}{3}\sqrt{2g\left(H_0 - \frac{2H_0}{3}\right)} = \sqrt{g}\, l_0 \left(\frac{2H_0}{3}\right)^{\frac{3}{2}} \quad [1.343]$$

The "unit" flow rate of the diversion will hence be:

$$q_{vf}(\delta) = = \sqrt{g}\, l_0 \left(\frac{2H_0}{3}\right)^{\frac{3}{2}} = 1{,}93.10^{-2}\, m^3 s^{-1} = 1666\, m^3.day^{-1} \quad [1.344]$$

2.3) The illegal diversions lead to a daily loss of water of $36\,000 - 13\,000 = 23\,000\, m^3.day^{-1}$. Their number can therefore be estimated at:

$$n = \frac{23000}{1666} = 138 \quad [1.345]$$

This is practically the equivalent of three diversions per kilometer!

The solution brought by the two previous examples can be generalized to many practical problems. It is particularly worth noting the extension of the channel problem to the film problem, a thin flow on the wall, which is distinct from both the boundary layer of a flat plate because of the fluid discontinuity, and from the jet because of the constant thickness of the liquid layer in the flow.

In order to show how the above can be used, an additional example will be provided. The approach of this example extracted from daily life accidents is perfectly applicable to the industrial field, notably to chemical engineering.

EXAMPLE 1.9 (A household incident).–

It is recommended to first solve the previous example. While painting a staircase, a do-it-yourselfer spills the paint pot. The result is that a vertical film of liquid flows down the riser. The paint will be assimilated to a Newtonian fluid of viscosity $\mu = 1{,}8.10^{-3}\ Pl$ and density $\rho = 1000\,kg.m^{-3}$. The flow rate of the film is $q_v = 8{,}10^{-6}\,m^3.s^{-1}$. Its width is $l = 5\,cm$. What is the thickness δ, considered constant, of the falling film?

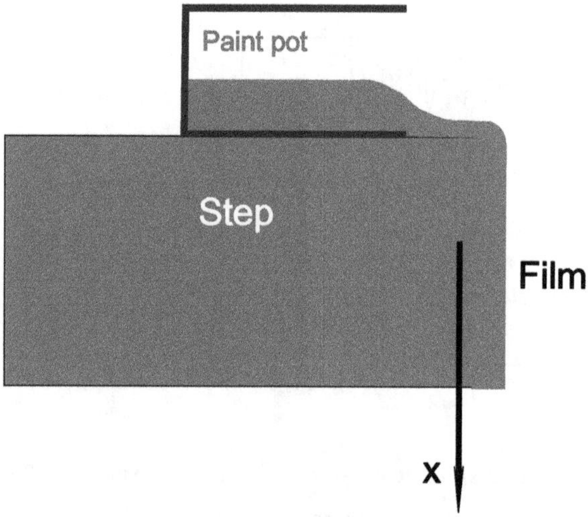

Figure 1.11. *A household incident*

Solution:

The results obtained in part 1 of the example entitled "Pont du Gard" can be readily applied here, noting that the angle $\alpha = \dfrac{\pi}{2}$. The equation will be written in a rigorously similar manner, and will yield the same result, with $\sin\alpha = 1$. The role of h is assumed by δ.

Hence:

$$q_v = \frac{gl\delta^3}{3\nu} \qquad [1.346]$$

$$\delta^3 = \frac{3*1{,}8.10^{-6}*8.10^{-4}}{9{,}81*5.10^{-2}} \qquad [1.347]$$

$$\delta = 2,065.10^{-3}\, m = 2,06\, mm \qquad [1.348]$$

This analysis is applicable in industrial settings, notably in the field of chemical engineering, in which the theory of the falling film is developed. In this context, the film facilitates the heat and mass transfers. In practice, this film is often located on an inner cylindrical surface of vertical axis. The previous result is still applicable, provided the thickness of the film remains lower than the radius of the tube.

2

One-dimensional Compressible Flows: Fully Reversible Flows

2.1. Introduction

The field of compressible flows is very broad. It is of practical interest in many situations: supersonic or hypersonic propulsion, turbomachinery. Acoustics is also covered in this chapter.

The scope here will be limited to one-dimensional flows under reversible adiabatic conditions (flow through a pipe of variable cross-section) and irreversible adiabatic conditions (one-dimensional shock wave).

This chapter is dedicated to *fully adiabatic and reversible flows*. Our focus will be on the nozzle problem.

In real-time applications, these restrictive hypotheses show their limitations immediately. The concept of shock will then be required. The next chapter will be dedicated to the brief study of *straight shock waves*.

Furthermore, in the same chapter, we shall see how a *combination of these two types of phenomenon* facilitates the treatment of situations that are closer to reality.

2.2. One-dimensional adiabatic and reversible flows

2.2.1. *Hypotheses adopted*

The problem examined here is that of the flow of a compressible gas through a piping system of variable section. Relative to the axis of reference Ox (axis of the pipe), the flow satisfies the following hypotheses:

a) The pipe has a known section, $S(x)$, which varies slowly with x. Therefore, the derivative $\frac{dS}{dx}$ is relatively low.

b) The flow is one dimensional.

Therefore, the velocity has a unique component V along Ox. The effects of an eventual radial component during the changes of section are neglected. This is consistent with the hypothesis of low $\frac{dS}{dx}$.

The temperature T of the fluid and its local density ρ depend exclusively on x.

In this chapter, the flow is assimilated to a series of "slices" of very small thickness $d\xi$ moving along the pipe with a velocity that takes the value $V(x)$ for the elevation x. Similarly to $V(x)$, each of these slices has a uniform temperature $T(x)$ and a density $\rho(x)$.

From an Eulerian perspective, each section of abscissa x is crossed at each moment of time by a slice of velocity $V(x)$, temperature $T(x)$ and density $\rho(x)$.

From a Lagrangian perspective, the observed slice "crosses" in its motion various sections of abscissa x and in each of them it "takes" the velocity, temperature and density "assigned" to that section according to the "Eulerian" description.

In this Lagrangian motion, a "slice" of fluid remains normal to the axis of the pipe, and undergoes variations in velocity, temperature and density. Therefore, it undergoes accelerations and variations in thermodynamic state. Mechanics and thermodynamics are coupled in these problems.

NOTE.– This is quite a common fact. Subsequent to the continuity and impulse relations, the energy equation is written, which is nothing other than an expression of the first law of thermodynamics. Although beyond the scope of this work, the study of convective transfers is such an example.

c) During its motion, the thermodynamic transformations undergone by this slice are assumed adiabatic and reversible. They are sufficiently rapid for the conductive transfer to neighboring slices and to the wall (consider heat-insulating, for the sake of simplicity) to be neglected, and sufficiently slow for the reversibility to be maintained. This amounts to considering that this is an "ideal" model.

d) The gas is inviscid and considered thermodynamically perfect.

2.2.2. *Writing the laws*

Writing the laws of continuity and dynamics and the first law of thermodynamics leads to three fundamental equations. The third equation introduces the specific enthalpy $h(x)$ of the gas.

The continuity equation:

$$\frac{d}{dx}\rho SV = 0 \qquad [2.1]$$

The fundamental law of dynamics:

$$\frac{dV^2}{2} + \frac{dp}{\rho} = 0 \qquad [2.2]$$

The first law of thermodynamics:

$$\frac{dV^2}{2} + dh = 0 \qquad [2.3]$$

The differentials represent the variations observed upon a displacement dx along the pipe.

2.2.3. *Other useful relations*

a) The mass flow rate q_m in the pipe can be written as:

$$q_m = \rho S V \qquad [2.4]$$

The conservation of this flow rate leads to equation [2.1].

b) The equation of state for perfect gases such as those recalled in Chapter 1 of [LED 17] (equations [1.34]–[1.38]) is:

$$\frac{p}{\rho} = rT \tag{2.5}$$

with

$$r = \frac{R}{M} \tag{2.6}$$

M is the molar mass of the gas. In contrast with R, r is not a universal gas constant, as it depends on the gas being considered.

c) According to Mayer's relation:

$$c_P - c_V = r \tag{2.7}$$

where c_P and c_V are specific heat capacities at constant pressure and constant volume, respectively, the ratio of which is:

$$\gamma = \frac{c_P}{c_V} \tag{2.8}$$

NOTE.– γ is often named "the" polytropic exponent of the gas. A central notion in thermodynamics, this parameter is equally important in what follows.

The definite article "the" is in fact inappropriate. From a formal point of view, during a polytropic process, the product p V^n is constant. γ is just a particular case of n.

d) Celerity of sound.

An important parameter is the celerity of sound $a(x)$ in a slice. This celerity is a local value related to the properties of the slice for a given abscissa.

The study of sound wave propagation is in fact a chapter of the physics of unsteady compressible flows. Within the framework of the *acoustic approximation*, the pressure, velocity and density disturbances caused by the passage of the wave are smaller than the static values of the variables (p, V, ρ). This approximation defines linear acoustics.

Shock waves do not obey the acoustic approximation. They are in the field of nonlinear acoustics, which explains the fact that their propagation celerity is different from the one calculated here.

In acoustics, it is shown that:

$$a = \frac{dp}{d\rho} \tag{2.9}$$

This ratio depends on the type of thermodynamic process that causes the variation in these two parameters during wave passage. If the process takes place when the wave passes and is assumed to be reversible adiabatic (isentropic), the celerity of sound in a point of the pipe will be given by:

$$a = \left.\frac{dp}{d\rho}\right|_S = \sqrt{\gamma r T} \tag{2.10}$$

e) Expression of the specific enthalpy h.

The specific enthalpy h of a perfect gas is defined by:

$$h = u + \frac{p}{\rho} \tag{2.11}$$

where u is the specific internal energy of the gas. It is known that for a perfect gas, we have:

$$u = c_V\ T \tag{2.12}$$

NOTE.– State functions, such as internal energy and enthalpy, are defined up to a constant. We are adopting here a frequently used convention, according to which the internal energy is null at absolute zero ($T = 0$). According to this same convention, the enthalpy is also null at the same absolute zero.

Consequently, several relations can be written as:

$$h = u + \frac{p}{\rho} = c_V\ T + rT = c_P\ T \tag{2.13}$$

These can be developed into:

$$h = c_p\, T = \frac{c_P}{r}\frac{p}{\rho} = \frac{c_P}{c_P - c_V}\frac{p}{\rho} = \frac{\gamma}{\gamma-1}\frac{p}{\rho} = \frac{a^2}{\gamma-1} \qquad [2.14]$$

f) NOTE.– a consequence of the reversible adiabatic hypothesis.

During a reversible adiabatic process of a perfect gas:

$$\frac{p}{\rho^\gamma} = Cst. \qquad [2.15]$$

When applied to two neighboring sections of a pipe through which there is a reversible adiabatic flow, this relation is written as a differential expression:

$$d\frac{p}{\rho^\gamma} = 0. \qquad [2.16]$$

After several manipulations, we can note that:

$$d\left(\frac{p}{\rho}\right) = \frac{dp}{\rho} - p\frac{d\rho}{\rho^2} \qquad [2.17]$$

$$d\left(\frac{p}{\rho^\gamma}\right) = \frac{dp}{\rho^\gamma} - \gamma p\frac{d\rho}{\rho^{\gamma+1}} = 0 \qquad [2.18]$$

From [2.18], it can readily be deduced:

$$d\rho = \frac{\rho\, dp}{\gamma p} \qquad [2.19]$$

which we combine with [2.17]:

$$d\left(\frac{p}{\rho}\right) = \frac{dp}{\rho} - p\frac{\rho\, dp}{\gamma p \rho^2} = \frac{dp}{\rho}\left(1 - \frac{1}{\gamma}\right) \qquad [2.20]$$

and it finally yields:

$$\frac{dp}{\rho} = \frac{\gamma}{\gamma-1}d\left(\frac{p}{\rho}\right) \qquad [2.21]$$

NOTE.– The appendix reminds us that equation [2.21] is a particular case of barotropic fluid.

The conservation of total enthalpy along the pipe (equation [2.3]) can be rewritten as:

$$\frac{dV^2}{2}+dh=\frac{dV^2}{2}+\frac{\gamma}{\gamma-1}d\left(\frac{p}{\rho}\right)=0 \qquad [2.22]$$

Using [2.21], we obtain:

$$\frac{dV^2}{2}+\frac{dp}{\rho}==0 \qquad [2.23]$$

which is similar to [2.2], expression of the fundamental law of dynamics.

This does not mean that equations [2.2] and [2.3] do not express different physical laws. It is because we have added an additional relation (through the hypothesis of reversible adiabatic process in section 2.2.1.c) that these two equations merge into one another.

NOTE.– The coincidence of these two equations is physically comprehensible: the fundamental law of dynamics entails a conservation of mechanical energy. The conservation of total enthalpy is a form of the first law of thermodynamics that conserves the sum of the mechanical energy (work) and thermal energy (heat).

In a reversible adiabatic process, there is neither heat loss (adiabatic character) nor internal transformation of work into heat (reversibility excludes viscous friction). Mechanical energy is therefore conserved.

g) The Mach number flow regimes.

In the treatment of high-velocity compressible flows, a dimensionless number plays a central role. Mach number is thus defined as the ratio of the velocity of a slice $V(x)$ to the celerity of sound $a(x)$ in that slice. Let us note that this celerity is a local value resulting from the local value of gas velocity and its local thermodynamic state:

$$M=\frac{V}{a}=\frac{V(x)}{a(x)} \qquad [2.24]$$

When comparing the material velocity of the fluid and the wave propagation celerity, three flow regimes can be evidenced:

The subsonic regime where $V < a$, therefore $M < 1$ [2.25]

The supersonic regime where $V > a$, therefore $M > 1$ [2.26]

The transonic (or sonic) regime where $V = a$, therefore $M = 1$ [2.27]

h) Generating state.

The compressible flow in a pipe has, in practice, a wide variety of origins: combustion, vapor generator, etc.

Various conclusions result from the possible manipulations of these equations.

A particular case arises when the piping system is connected to a "reservoir", that is, to a volume with sufficiently large section for the velocity to be null in it. Even though the flow is most often generated by a complex system, it can be considered identical to a flow emerging from a reservoir.

The state in this reservoir is such that, by noting the properties of it with a subscript "0", we have:

$$h_0 = h(x) + \frac{V^2(x)}{2} \quad [2.28]$$

$$h_0 = \frac{a_0^2}{\gamma - 1} = \frac{a^2(x)}{\gamma - 1} + \frac{V^2(x)}{2} \quad [2.29]$$

$$\frac{p_0}{\rho_0} = r T_0 \; h_0 = \frac{\gamma}{\gamma - 1}\frac{p_0}{\rho_0} = \frac{\gamma}{\gamma - 1}\frac{p}{\rho} + \frac{V^2(x)}{2} \quad [2.30]$$

$$a_0 = \sqrt{\gamma r T_0} \quad [2.31]$$

Knowing V, p and ρ in an arbitrary section, V_0, p_0 and ρ_0 can be determined without ambiguity.

i) Maximum velocity.

Given that we have now defined and we know the generating state, let us imagine that we had the possibility to obtain a flow state in which the entire potential energy is transformed into kinetic energy. In this section, the pressure would be p_0 and the value of velocity would be V_{max} so that:

$$h_0 = \frac{\gamma}{\gamma-1}\frac{p_0}{\rho_0} = \frac{2a_0^2}{\gamma-1} = \frac{V_{max}^2}{2} \quad [2.32]$$

$$V_{max} = \sqrt{\frac{2\gamma}{\gamma-1}\frac{p_0}{\rho_0}} = a_0\sqrt{\frac{2}{\gamma-1}} \quad [2.33]$$

V_{max} cannot be physically reached, but it is a characteristic of the flow. It is directly related to the generating state.

j) Critical velocity.

A critical velocity V_C, which corresponds to a specific section (at the throat, as we shall see), is also defined as: $V_C = a$. This particular celerity of sound is named critical celerity and denoted by a_C.

We deduce that $V_C = a_C$ [2.34]

$$h_0 = h + \frac{V^2}{2} = \frac{2a_C^2}{\gamma-1} + \frac{V^2}{2} \quad [2.35]$$

$$h_0 = \frac{2a_0^2}{\gamma-1} = \frac{2a_C^2}{\gamma-1} + \frac{a_C^2}{2} \quad [2.36]$$

$$V_C = a_C = a_0\sqrt{\frac{2}{\gamma+1}} \quad [2.37]$$

2.2.4. *Fundamental relations*

As a result of sometimes heavy mathematical manipulations, a certain number of fundamental relations and practical applications can be deduced.

a) The relation between velocity and local pressure

$$V = \sqrt{\frac{2\gamma r T}{\gamma-1}\left[1-\left(\frac{p}{p_0}\right)^{\frac{\gamma-1}{\gamma}}\right]} = \sqrt{\frac{2a^2}{\gamma-1}\left[1-\left(\frac{p}{p_0}\right)^{\frac{\gamma-1}{\gamma}}\right]} \quad [2.38]$$

b) The Hugoniot relation

$$\frac{dS}{S}+\frac{dV}{V}\left(1-M^2\right)=0 \qquad [2.39]$$

This very important relation shows how the Mach number acts in various flow regimes.

In subsonic regime, the Mach number is less than 1:

$$M<1, \text{ hence } V<a \text{ then } \left(1-M^2\right)>0 \qquad [2.40]$$

which leads to $\frac{dS}{S}$ and $\frac{dV}{V}$ having opposite signs.

In supersonic regime, the Mach number is higher than 1:

$$M>1, \text{ hence } V>a \text{ then } \left(1-M^2\right)<0 \qquad [2.41]$$

which leads to $\frac{dS}{S}$ and $\frac{dV}{V}$ having the same sign.

In transonic regime, the Mach number is equal to 1:

$$M=1, \text{ hence } V=a \text{ then } \left(1-M^2\right)=0 \qquad [2.42]$$

which leads to $\frac{dS}{S}$ being null.

The previous remarks suggest that a profound distinction separates compressible flows, namely subsonic and supersonic compressible flows.

In subsonic regime, $\frac{dS}{S}$ *and* $\frac{dV}{V}$ *have opposite signs*. Velocity increases when section area decreases. This phenomenon is qualitatively observed with an incompressible fluid.

NOTE.– It is worth noting that the relative variations of V and S are not identical to the ones for the incompressible flow, since ρ also varies from one section to another.

In supersonic regime, $\frac{dS}{S}$ *and* $\frac{dV}{V}$ *have the same sign.* At a first glance, this baffles intuition.

NOTE.– Velocity increases when section decreases. This cannot be comprehended unless one is aware of the fact that the thermodynamic state changes along the flow and the density consequently varies.

In subsonic regime, $\frac{dS}{S}$ *is null. An extremum is reached* in the pipe.

An important consequence of this Hugoniot relation:

Any regime change (transition from subsonic to supersonic flow or *vice versa*) is possible only for an extremum of the pipe section area.

In effect, this transition involves a Mach number equal to 1. This extremum could *a priori* be a maximum (bulge) or a minimum (throat). A simple reasoning shows that this extremum can only be a minimum. Let us assume a transition takes place at a bulge of the pipe. In a subsonic regime, upstream of this maximum on the pipe, flow velocity would decrease upon approaching it and the Mach number would diminish. It would be impossible to reach 1 at this maximum of the section area. A similar reasoning can be applied to a supersonic flow.

The transition from subsonic to supersonic flow is therefore possible only at a throat. At this throat, the flow will have to be sonic (or transonic).

NOTE.– It is worth recalling that all this relies on the hypothesis of flow reversibility.

We shall see that there are *local irreversibilities* in which the transition from supersonic to sonic regime is possible outside a throat (shock waves).

c) Relations between the state variables and the Mach number.

The flow variables $p(x)$, $\rho(x)$ and $T(x)$ can also be related to the local Mach number. Using the generating state, we obtain dimensionless expressions. We shall extensively use this remark by resorting to tables.

Hence, for pressure $p(x)$, we have:

$$\frac{p}{p_0} = \left(1 + \frac{\gamma - 1}{2} M^2\right)^{\frac{\gamma}{1-\gamma}} \quad [2.43]$$

Since the flow is adiabatic and reversible, we know that:

$$\frac{p}{\rho_0^{\gamma}} = \frac{p}{\rho^{\gamma}} \quad [2.44]$$

which we use to write the following relations:

$$\frac{\rho}{\rho_0} = \left(\frac{p}{p_0}\right)^{\frac{1}{\gamma}} = \left(1 + \frac{\gamma - 1}{2} M^2\right)^{\frac{1}{1-\gamma}} \quad [2.45]$$

$$\frac{T}{T_0} = \frac{p}{p_0}\frac{\rho_0}{\rho} = \left(1 + \frac{\gamma - 1}{2} M^2\right)^{-1} \quad [2.46]$$

These expressions are used when setting up tables.

2.2.5. Calculation of flow rate in a piping system

Determining the mass flow rate is an essential calculation in any compressible flow-related problem.

a) The calculation starts from the relation between this flow rate and the generating state. Either real or imaginary, this generating state sets the mass flow rate in the piping system.

Using the generating state, and based on the previous expressions (notably [2.39]), we can show after several mathematical manipulations that this flow rate can be calculated in each section S starting from the following expression, which requires knowledge of the generating state:

$$q_m = \rho_0 V_{\max} S Y\left(\frac{p}{p_0}\right) \quad [2.47]$$

where p and p_0 are the pressure values corresponding to the section S and the generating state, respectively.

This function Y can be written as:

$$Y\left(\frac{p}{p_0}\right)=\left(\frac{p}{p_0}\right)^{\frac{1}{\gamma}}\left[1-\left(\frac{p}{p_0}\right)^{\frac{\gamma-1}{\gamma}}\right]^{\frac{1}{2}} \quad [2.48]$$

It is a bell-shaped function. It reaches a maximum Y_C for a pressure value p_C, according to:

$$\frac{p_c}{p_0}=\left(\frac{2}{\gamma+1}\right)^{\frac{\gamma}{\gamma-1}} \quad [2.49]$$

$$Y_C\left(\frac{p_c}{p_0}\right)=\left(\frac{2}{\gamma+1}\right)^{\frac{1}{\gamma-1}}\frac{\gamma-1}{\gamma+1} \quad [2.50]$$

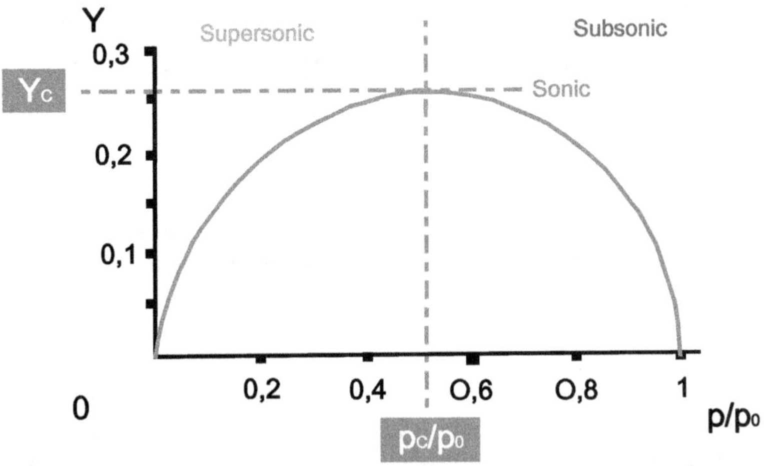

Figure 2.1. *Graphical representation of the function* $Y\left(\frac{p}{p_0}\right)$

b) It can be noted that ρ_0 and V_{max} are characteristic of the generating state. Therefore, the flow rate conservation requires the product SY to be constant throughout the piping system.

Two values of S are remarkable:

– the throat section S_C;

– the piping system outlet section S_S.

We shall therefore recall that:

$$SY = S_C Y_C = S_s Y_s \qquad [2.51]$$

and this yields $\dfrac{Y}{Y_C} = \dfrac{S_C}{S}$ [2.52]

c) Maximum flow rate.

Let us note at this point that in formula [2.47], the generating state sets ρ_0 and V_{max}. The section at the throat S_C is set by construction. Let us calculate the flow rate at the throat. In a general case, Y can take any value between 0 and Y_C:

$$q_m = \rho_0 V_{max} S_C Y \qquad [2.53]$$

Let us assume that the sonic regime is obtained at the throat. Then, Y reaches its maximum possible value denoted by Y_C. The flow rate becomes:

$$q_m = \rho_0 V_{max} S_C Y_C \qquad [2.54]$$

It can be seen that for a given generating state and a fixed diameter of the throat, the flow rate in a piping system cannot exceed the value obtained when a sonic regime is established at the throat. When the flow is "frozen", it is said to be "choked".

$\frac{p}{p_0}$ is expressed as a function of the Mach number, and Y, which depends on it, can be related to the Mach number. In the end, it is possible to express $\frac{S}{S_C}$ as a function of the Mach number for a section S:

$$\frac{S}{S_C} = \frac{1}{M}\left[\frac{2}{\gamma+1}\left(\frac{\gamma-1}{2}M^2+1\right)\right]^{\frac{1}{2}+\frac{1}{\gamma-1}} \quad [2.55]$$

d) Physical determination of the flow rate.

From a physical perspective, a flow is *a priori* determined by the generating state and the upstream pressure. As we shall see, there are situations in which the flow rate is determined only by the generating flow. The generating state can be real (reservoir) or virtual. Knowing this flow in one section, it is possible to determine the generating state. In particular, its pressure p_0 will be known. To illustrate the flow rate determination problem, we shall use a common pipe model, namely a de Laval nozzle. This nozzle consists of a converging section, a throat and a diverging section.

NOTE.– In an ideal configuration, we assume that the converging part is connected by an "infinite" section to the reservoir in which there is a generating state. When there is a real reservoir, the connection section is finite, but it is sufficiently large to allow us to assume that fluid velocity in the reservoir is null (which is essentially the definition of the generating state). If the generating state is virtual, so is the "infinite" connection to the reservoir.

The outlet pressure of the upstream nozzle is decreased starting from the value of the generating flow p_0, and the evolution of the mass flow rate in this nozzle is calculated. Our reasoning is based on formula [2.56]:

$$q_m = \rho_0 V_{max} SY\left(\frac{p}{p_0}\right) \quad [2.56]$$

The representative curve of Y in Figure 2.1 can be divided into two parts: an increasing part on the left, between $\frac{p}{p_0} = 0$ and $\frac{p}{p_0} = \frac{p_c}{p_0}$

and a decreasing part on the right, between $\frac{p}{p_0} = \frac{p_c}{p_0}$ and $\frac{p}{p_0} = 1$

On the part on the right-hand-side, for a given section, the pressure increases when the flow rate decreases. This behavior characterizes a subsonic flow. On the left-hand-side part, pressure increases when the flow rate increases. This is how a supersonic flow behaves.

It is worth noting that the same flow rate can lead to two representative points, on each part of the curve.

In summary, the left part refers to supersonic flows and the right part refers to subsonic flows. The maximum corresponds to a sonic flow.

2.2.6. *De Laval nozzle*

As already noted, de Laval nozzle consists of a converging pipe, for which $\frac{dS}{dx} < 0$, connected by a throat, for which $\frac{dS}{dx} = 0$ (extremum), to a diverging pipe, for which $\frac{dS}{dx} > 0$. It is a proper model for understanding what determines the flow rate of a compressible flow through a pipe. Let us denote the outlet section pressure by p_S. We recall that when the flow is sonic at the throat, the pressure at the throat is equal to the critical pressure p_C.

Let us note that, as long as the hypotheses of an isentropic flow in the nozzle are assumed valid, for a given generating flow, only two outlet section pressures, which will be denoted by p_{s1} and p_{s2}, authorize a sonic flow at the throat. They are such that:

$$Y_s\left(\frac{p_{S1}}{p_0}\right) = \frac{S_C Y_C}{S_S} \quad [2.57]$$

$$Y_s\left(\frac{p_{S2}}{p_0}\right) = \frac{S_C Y_C}{S_S} \quad [2.58]$$

These two distinct values of Y_S are located on each of the parts of the curve $Y = Y\left(\frac{p}{p_0}\right)$. The two corresponding flows will therefore be subsonic and supersonic, respectively.

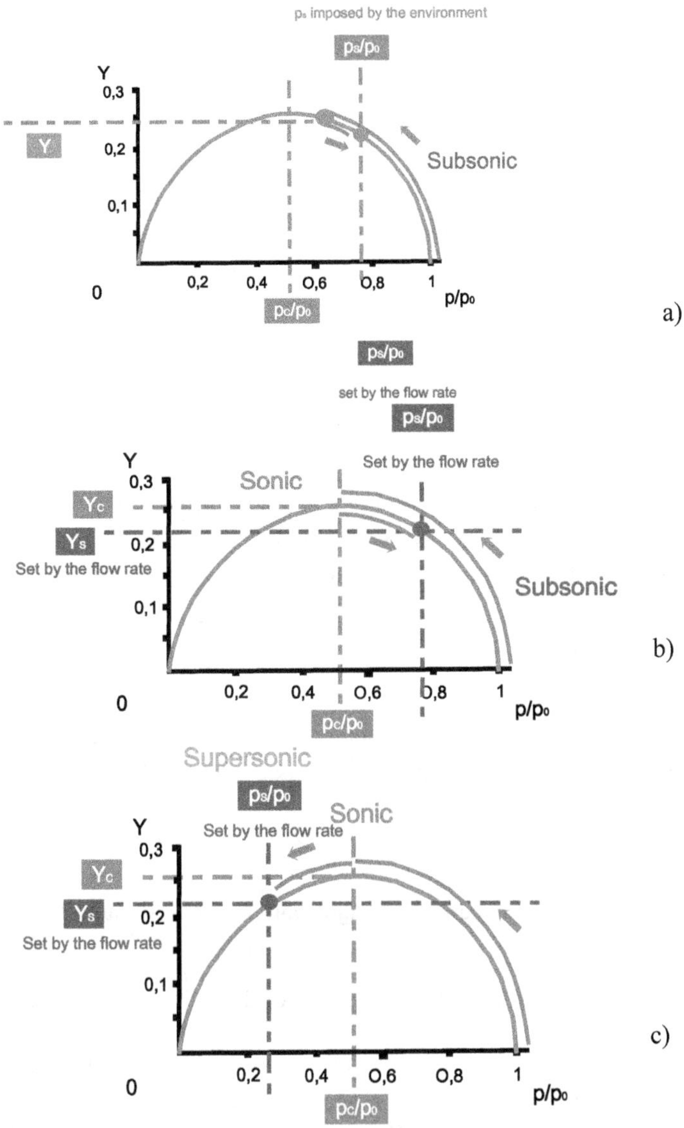

Figure 2.2. *Displacement of the representative point of the flow on the* $Y\left(\frac{p}{p_0}\right)$ *curve.*

a) Subsonic flow in the converging and diverging parts. b) Sonic flow at the throat, subsonic flow in the converging and diverging parts. c) Sonic flow at the throat, subsonic flow in the converging part and supersonic flow in the diverging part

When setting an upstream generating state, downstream pressure is decreased starting from the pressure p_0 of the generating state. Let us observe what happens at the throat, as for any other point in the nozzle, the observed flow at representative point on the $Y\left(\frac{p}{p_0}\right)$ curve.

As long as $p_S > p_{S1}$, the flow rate increases inversely to p_{S1}. It is subsonic in the converging and diverging parts, as well as at the throat. The representative point of pressure $p(x)$ covers a sector of the (subsonic) right part of $Y = Y\left(\frac{p}{p_0}\right)$ starting from $\frac{p}{p_0} = 1$ and comes back to $\frac{p}{p_0} = \frac{p_S}{p_0}$.

When $p_S = p_{S1}$, the flow is subsonic in the converging and diverging parts and sonic at the throat (which is set by the flow rate). The flow is choked. The representative point of the pressure $p(x)$ covers a sector of the right part of $Y = Y\left(\frac{p}{p_0}\right)$ starting from $\frac{p}{p_0} = 1$ and comes back to $\frac{p}{p_0} = \frac{p_{S1}}{p_0}$.

When $p_S = p_{S2} < p_{S1}$, the flow is subsonic in the converging parts, sonic at the throat (which is set by the flow rate) and supersonic in the diverging part. In effect, $\frac{p}{p_0} < \frac{p_C}{p_0}$ in this section. The nozzle is choked. The representative point of the pressure $p(x)$ covers a sector of the right part $Y = Y\left(\frac{p}{p_0}\right)$ up to $\frac{p_C}{p_0}$ and goes back to $\frac{p}{p_0} = \frac{p_{S2}}{p_0}$ on the right-hand-side part of $Y = Y\left(\frac{p}{p_0}\right)$. The values p_{S1} and p_{S2} of the outlet pressure relate to the only two cases of choked nozzle for which the flow, either sonic or supersonic, is isentropic throughout the nozzle. In this case, the nozzle is said to be "adapted". These conditions are not convenient in practice. Imagine a propeller with a nozzle operating with only two outlet pressures.

It is worth questioning what happens when the outlet pressure is less than p_{S1} and differs from p_{S2}. We shall see that the reversibility hypothesis will have to be abandoned in one point of the nozzle: this will lead to the emergence of the shock wave. This important phenomenon will be explored in 3.2.

Furthermore, in 3.3, we shall see how reversible flows and shock wave can combine in a nozzle. To start with, in section 2.3, we focus on examples of reversible flows.

2.3. Applications. Reversible adiabatic flows

The preceding formulae show that the field of compressible flows can lead to complex calculations, even for a simplified situation. Thanks to numerical tools, the calculation phase can be rendered less cumbersome. We shall indirectly use this numerical tool when resorting to multiple entry tables, which significantly simplify the manipulation of the relations established above.

As presented in the appendix to this book, the reader can easily build these tables. It has been considered convenient to take the Mach number as an entry data in these tables.

At the end of this book, there is a table related to isentropic flows. From the same perspective, the following chapter ends with a table related to straight shocks. The tables given in this book have been established for $\gamma = 1.4$, and this value was chosen for treating all the following examples. The reader can refer to tables which have already been calculated when solving the problems. This "standard" value of 1.4 is valid for the air and for several other gases with a satisfactory approximation degree. Moreover, gases will be assumed "weightless", meaning that the effect of volume forces will be neglected, similarly to the equation given in [2.2]. Although it is a mixture, air will be assimilated to a body of molar mass $M = 29\,g$.

EXAMPLE 2.1 (Compressed-air duster for electronic devices).–

A can designed to generate an air jet that removes dust from inside computers and other electronic devices consists essentially of a cylinder of inner diameter $D = 6\,cm$ and height $H = 20\,cm$.

The inner pressure in the can is maintained at a constant value $p_i = 2{,}5\,bar$ throughout any use, based on a method that will not be described here. The air exits through a valve, the structure of which will not be described here. Let us simply note that the air exits through a tube that opens on its smallest section, with a diameter $d = 0.5\,mm$. The temperature is at any point constant and equal to $21°C$. Let us calculate the can emptying time, based on two series of hypotheses.

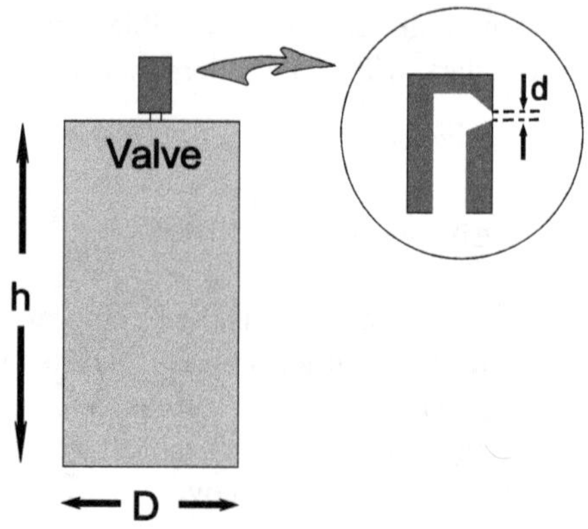

Figure 2.3. *The compressed-air generator and its valve*

Hypothesis 1: The air is incompressible and weightless. It retains the same density as in the can, even in the vicinity of its valve outlet. The jet is practically cylindrical at the aperture.

Hypothesis 2: Air compressibility is taken into account.

1) Calculation according to hypothesis 1

1.1) What is the velocity of gas when it exits the aperture? What is its mass flow rate?

1.2) What is your estimation of the duration τ_1 of the can?

2) Calculation according to hypothesis 2

2.1) Based on the calculation of the critical pressure, show that the flow is sonic at the exit aperture.

2.2) Calculate the emptying mass flow rate.

2.3) Deduce from it the real operating time τ_2 of the can. To what extent is hypothesis 1 reasonable?

Solution:

The first parameter to search for in nozzle-related problems is the generating state. It can be either given, which is the present case, or deduced from the data related to the flow and gas state when this is perfectly known in a section of the pipe.

The generating state is provided here by:

the constant pressure in the can, $p_0 = 2.5\,bar = 2,5.10^5\ Pa$
the temperature $T_0 = 21C = 294\,K$.

The density ρ_0 can be deduced from the equation of state for perfect gases:

$$\frac{p_0}{\rho_0} = r\,T_0 \qquad [2.59]$$

$$r = \frac{R}{M} = \frac{8,31}{29.10^{-3}} = 286,5 \qquad [2.60]$$

$$\rho_0 = \frac{p_0}{r\,T_0} = \frac{2,5.10^5}{285,5.293} = 2,99\,kg.m^{-3} \qquad [2.61]$$

Let us note that:

T_0 must be expressed in Kelvin

r is not a constant, but it depends on the nature of the gas (i.e. on its molar mass)

The molar mass M, generally given in grams, must be expressed in kilograms in all formulae. Neglecting this remark results in values of r of the order of 0.3 while the order of magnitude of this parameter (except for hydrogen and heavy gases) is rather 300. Taking this remark into consideration would avoid the reader making an error encountered in many papers.

Two important parameters can be readily calculated:

The celerity of sound in the reservoir:

$$a_0 = \sqrt{\gamma\,r\,T_0} = 343\,m.s^{-1} \qquad [2.62]$$

The theoretical maximum velocity of any flow coming out of this reservoir:

$$V_{\max} = a_0\sqrt{\frac{2}{\gamma - 1}} = 767,9\,m.s^{-1} \qquad [2.63]$$

In both hypotheses, the area of the aperture through which the air flows is:

$$s = \pi \frac{d^2}{4} = 1{,}96.10^{-7}\ m^2 \qquad [2.64]$$

Considering the set of questions posed, it is interesting to calculate the mass of air m contained in the can before use.

The volume of the can is:

$$V_{OL} = \pi \frac{D^2}{4} h = 5{,}65.10^{-4}\ m^3 \qquad [2.65]$$

$$m = \rho_0 V_{OL} = 1{,}68.10^{-3}\ kg \qquad [2.66]$$

1) The gas is a perfect fluid.

1.1) The outlet velocity can be determined by applying Bernoulli's theorem on a flow line between a point in the reservoir (where velocity is null) and the exit section. The density is considered constant and equal to ρ_0. The gas being "weightless", the terms in $p_i = 2{,}5\,bar$ will not be taken into account.

$$\rho_0 \frac{(0)^2}{2} + p_0 = \rho_0 \frac{V_S^{\ 2}}{2} + p_a \qquad [2.67]$$

where V_S is the outlet velocity at the valve and p_a is the atmospheric pressure.

The classical expression of the velocity when exiting a reservoir is:

$$V_S = \sqrt{2 \frac{p_0 - p_a}{\rho_0}} \qquad [2.68]$$

$$V_S = 318\, m.s^{-1} \qquad [2.69]$$

The mass flow rate can be deduced:

$$q_{m1} = \rho_0 V\, s = 1{,}85.10^{-4}\ kg.m^{-3} \qquad [2.70]$$

1.2) The time duration required to empty the mass $m = 1.68.10^{-3}\,kg$ contained in the can is:

$$\tau_1 = \frac{m}{q_{m1}} = 9,08\,s \tag{2.71}$$

Let us note at this stage that the incompressibility hypothesis is quite unrealistic for a flow exiting at a Mach number M_1 close to 1:

$$a_0 = 343\,m.s^{-1};\ M_1 = \frac{318}{343} = 0,93 \tag{2.72}$$

2) Gas compressibility is taken into account.

2.1) The critical pressure is a pressure in a section where fluid velocity is equal to the celerity of sound, hence a Mach number equal to 1.

It can be calculated directly using the formula recalled in [2.1]:

$$\frac{p_c}{p_0} = \left(\frac{2}{\gamma+1}\right)^{\frac{\gamma}{\gamma-1}} \tag{2.73}$$

If isentropic flow tables are available, we can readily calculate its ratio to p_0 on the line where the Mach number is 1. We therefore have:

$$M = 1 \Rightarrow \frac{p}{p_0} = \frac{p_C}{p_0} = 0,528 \tag{2.74}$$

$$p_C = 0,528 * 2,5.10^5 = 1,32.10^5 \tag{2.75}$$

This pressure $p_C = 1.32 bar$ is higher than the atmospheric pressure of 1 bar. The valve will be blocked.

We have already explained the blocking phenomenon in the case of de Laval nozzle. We can resume here an analogous reasoning for what we here name simply a converging nozzle.

Let us resume a reasoning used in 2.2.5 with de Laval nozzle: let us imagine that we place the can in a chamber where pressure decreases from p_0. The flow in the valve will be subsonic, and hence the pressure in the chamber will be higher than critical pressure. For critical pressure at the throat, the regime is sonic. The flow rate at the throat is calculated starting from the critical value of Y. For a unity Mach number, the table indicates:

$$Y_C = 0{,}2588 \quad [2.76]$$

This value could have been retrieved using the direct formula:

$$Y_C\left(\frac{p_c}{p_0}\right) = \left(\frac{2}{\gamma+1}\right)^{\frac{1}{\gamma-1}} \frac{\gamma-1}{\gamma+1} \quad [2.77]$$

q_m is then calculated based on the expression:

$$q_m = \rho_0\, S V_{\max}\, Y_C \quad [2.78]$$

Y_C is the maximum possible value of Y. The flow rate in the valve cannot exceed q_m found in [2.77].

If we diminish the pressure in the chamber, the pressure at the throat will remain equal to p_C and the transition to the outside will not be possible through a reversible process. It is said that the nozzle is choked at its maximum flow rate, which can no longer depend on the upstream conditions.

This choking phenomenon facilitates the *flow rate control of a compressible fluid only based on upstream pressure*, independent of any downstream condition. This device is widely used, notably in laboratories, for example, to control experiments of combustion under vacuum. This device is known as *sonic throat.* The physically possible extremum of velocity $V_{\max}$ is:

$$V_{\max} = a_0\sqrt{\frac{2}{\gamma-1}} = 765\, m.s^{-1} \quad [2.79]$$

$$q_{m2} = \rho_0\, S V_{\max}\, Y_C = 2{,}97*1{,}96.10^{-7}*.767{,}9*0{,}2588 = 1{,}28.10^{-4}\, kg.s^{-1} \quad [2.80]$$

Hence, the can emptying time τ_2 is:

$$\tau_2 = \frac{m}{q_{m2}} = \frac{1{,}68.10^{-3}}{1{,}28.10^{-4}} = 13{,}1\, s \quad [2.81]$$

It can be seen that the incompressibility hypothesis yields a rough order of magnitude. The error leads to underestimate the time by 40%.

EXAMPLE 2.2 (Tire leak).–

The pressure in a new tire is 2,6 bar. Its air volume is 0,03 m^3. The air temperature, which is assumed constant throughout the problem, is 30°C. The atmospheric pressure is constant at 1 bar.

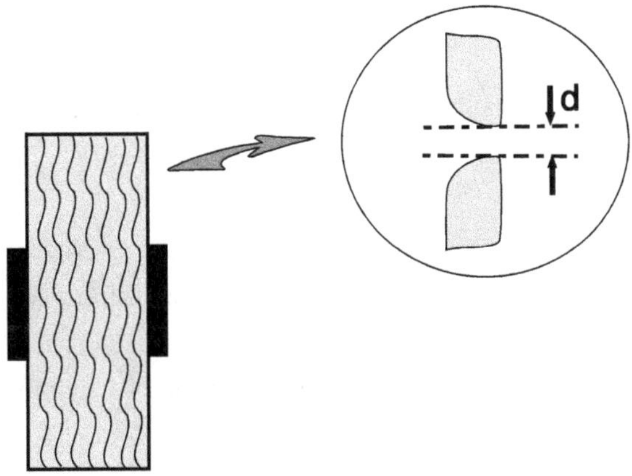

Figure 2.4. *The tire and its air leak*

When a tire is punctured, it deflates. The hole is assumed analogous to a converging nozzle with an outlet diameter of 1 mm. We shall study the variation in pressure in the tire between 2,6 and 1,9 bar. For this purpose, we shall assume that the inner volume of the tire does not vary. The polytropic exponent of air is 1,4. The gas is considered perfect.

1) Show that for the tire pressures considered, the air leakage flow rate depends only on the inner pressure of the tire.

2) Find the relation between the instantaneous leakage mass flow rate, q_m (t) and the density, then the tire pressure at a given moment t, p_0 (t). The process is assumed quasi-steady.

3) Deduce the differential equation satisfied by the mass of air contained by the tire. Then, deduce the differential equation satisfied by p_0 (t).

4) How long does it take for the tire pressure to decrease from $2{,}6\,bar$ to $1{,}9\,bar$?

Solution:

1) The tire emptying flow rate does not depend on the upstream pressure provided the sonic regime is "blocked" at the throat. The nozzle must be choked.

The tire pressure and the air temperature define the generating state:

$$p_0 = 2,6\,bar = 2,6.10^5\ Pa;\ T_0 = 30\,°C = 303\,K \qquad [2.82]$$

When the nozzle is choked, the ratio of generating state pressure p_C to the pressure at the throat p_0 is known; it depends on the value of γ. The ratio $\frac{p_C}{p_0}$ is 0,5283 for $\gamma = 1{,}4$.

This value can be calculated by the previously found formula, $\frac{p_c}{p_0} = \left(\frac{2}{\gamma+1}\right)^{\frac{\gamma}{\gamma-1}}$ or more simply in the table of adiabatic flows, in the $\frac{p}{p_0}$ column and on the line corresponding to $M = 1$.

The flow is sonic at the throat. The transition to the atmosphere outside the tire is an irreversible phenomenon. This can only be possible if the pressure p_c is at least equal to or higher than the atmospheric pressure $p_a = 10^5\ Pa$. For a polytropic exponent γ of 1,4, the nozzle will only be choked for tire pressures $p_C > p_a$:

$$p_C = 0,5283.p_0 > 10^5 \qquad [2.83]$$

$$\text{or } p_0 > \frac{10^5}{0,5283} = 1,89.10^5\ Pa \qquad [2.84]$$

This will be particularly verified by all the pressures between $2,6\,bar$ and $1,9\,bar$.

The previous calculation is rigorously valid only for permanent flow. In the hypothesis of a slowly varying flow, we shall opt for a quasi-steady reasoning. The relations between pressures and flow rates that are valid under steady regime are acceptable when the tire pressure varies slowly. The hypothesis can simply be validated in relation to the result obtained for pressure decrease time.

2) The mass flow rate (in this problem, an "instantaneous" flow rate) at the throat is written as:

$$q_m = \rho_0 V_{max} S_c Y_C \quad [2.85]$$

$$S_c = \pi \frac{d^2}{4} = 7{,}85.10^{-7}\ m^2 \quad [2.86]$$

For a more precise writing of q_m, several "time-dependent" characteristics of the generating state must be defined:

$$\rho_0 = \frac{p_0}{r T_0} = \alpha\, p_0 \quad [2.87]$$

$$r = \frac{R}{29.10^{-3}} = 286{,}7;\ T_0 = 303\, K;\ \alpha = 1{,}151.10^{-5}\ SI \quad [2.88]$$

$$V_{max} = \sqrt{\frac{2}{\gamma - 1}}\, a_0 \quad [2.89]$$

$$a_0 = \sqrt{\gamma r T_0} = 348{,}7\, ms^{-1} \quad [2.90]$$

$$V_{max} = 779{,}7\ ms^{-1} \quad [2.91]$$

Moreover, for $\gamma = 1.4$, the table for adiabatic flows indicates $Y_C = 0{,}2588$ on the line $M = 1$.

It follows:

$$q_m = \alpha\, p_0\, 779{,}7 * 7{,}85.10^{-7} * 0{,}2588 \quad [2.92]$$

$$q_m = \beta\, p_0;\ \beta = 1{,}82.10^{-9}\ SI \quad [2.93]$$

Here we resort to introducing constants α, β that ease the writing of the expressions and minimize the risk of error in their manipulation.

3) The variation of the air mass in the tire is given by the following differential equation accompanied by its boundary condition:

$$\frac{dm}{dt} = -q_m = -\beta p_0 \quad [2.94]$$

$$t = 0;\ m = m_{init} \quad [2.95]$$

To transform this into an equation in p_0, we must express p_0 as a function of *m*:

$$m = \rho_0 V_{OL} = 0,03\,\rho_0 = 0,03\,\alpha\,p_0 = 3,45.10^{-7}\ p_0 \quad [2.96]$$

$$0,03\,\alpha \frac{dp_0}{dt} = -\beta\,p_0 \quad [2.97]$$

$$\frac{dp_0}{dt} = -\delta\,p_0;\ \delta = \frac{\beta}{0,03\,\alpha} = 5,2.10^{-3} \quad [2.98]$$

The initial condition is given by:

$$t = 0;\ \ p_0 = p_{0init} = 2,6.10^5 \quad [2.99]$$

The resolution of this equation is elementary and classical:

$$\frac{dp_0}{p_0} = -\delta\,dt \quad [2.100]$$

$$Ln\,p_0 = -\delta t + Ln C \quad [2.101]$$

$$p_0 = C\,e^{-\delta t};\ t = 0;\ \ p_0 = p_{0init} \quad [2.102]$$

$$\frac{p_0}{p_{0init}} = e^{-\delta t} \quad [2.103]$$

When the pressure decreases from $2,6\,bar$ to $1,9\,bar$, t is calculated as:

$$\frac{dp_0}{p_0} = -\delta\,dt \quad [2.104]$$

$$Ln\, p_0 = -\delta t + Ln C \qquad [2.105]$$

$$p_0 = C\, e^{-\delta t}; t = 0;\ p_0 = p_{0init} \qquad [2.106]$$

$$t = -\frac{1}{\delta} Ln\left(\frac{p_0}{p_{0init}}\right) = -\frac{1}{5,2.10^{-3}} Ln\left(\frac{1.9}{2,6}\right) = 60,3\ s \qquad [2.107]$$

Therefore, the tire deflates within 1 min. Let us note that this may seem a short time, but the hole in the tire has a diameter of 1 mm, which is not negligible. As a general rule, when a sharp object (screw, nail) punctures a tire, it sticks into the air tube. This way, the air leakage hole is smaller, and the loss of pressure unfolds over a longer duration (several hours). The case treated here would rather relate to a tire punctured, for example, by a sharp stone.

EXAMPLE 2.3 (Vapor exhaust from a pressure cooker).–

A pressure cooker contains 500 g of water that will be assumed to be in a state of liquid–vapor equilibrium throughout the problem, at a temperature $T_{H2O} = 160\, C$. Saturated water vapor pressure at this temperature is $p_S = 6,17.10^5\ Pa$. At this pressure, vapor can still be considered a perfect gas.

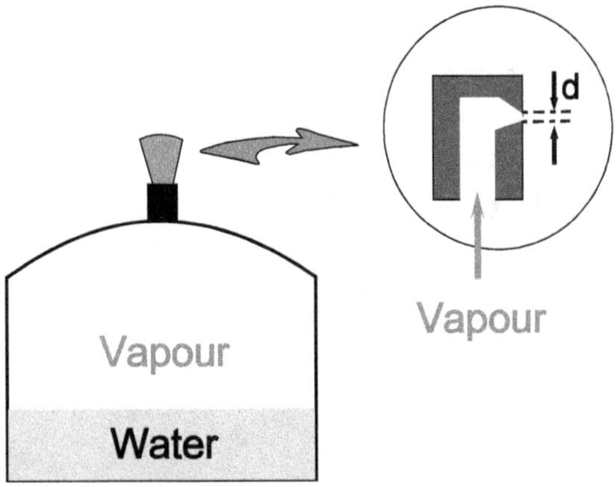

Figure 2.5. *Pressure cooker and its valve*

The valve of the pressure cooker is opened. This operation is assumed to be the equivalent of opening the tank to the outside environment through a converging nozzle having an outlet aperture of diameter $D = 2\,mm$. The atmospheric pressure in the kitchen is $p_a = 10^5\ Pa$. The polytropic exponent γ for water vapor is assumed to be 1,4.

1) Show that the sonic regime is choked in the nozzle.

2) Deduce the mass flow rate in this nozzle.

3) Is it possible to make an approximate estimation of how long it will take for the 500 g of water to be evacuated from the pressure cooker? The temperature in the pressure cooker is considered constant.

Solution:

1) Similarly to any problem of this type, we shall use the properties of the generating flow.

We know p_0 and T_0. The density will be calculated from the equation of state for perfect gases.

$$\frac{p_0}{\rho_0} = r\,T_0 \qquad [2.108]$$

$$p_0 = p_S = 6{,}17.10^5\ Pa;\ T_0 = 160 + 273 = 433\,K \qquad [2.109]$$

$$M_{H2O} = 18\,g;\ r = \frac{R}{M_{H2O}} = \frac{8{,}31}{18.10^{-3}} = 461{,}7 \qquad [2.110]$$

$$\rho_0 = \frac{p_0}{r\,T_0} = \frac{6{,}17.10^5}{461{,}7*433} = 3{,}09\,kg.m^{-3} \qquad [2.111]$$

Let us determine the celerity of sound and the enthalpy of the generating state, as well as the theoretical maximum velocity:

$$a_0 = \sqrt{\gamma r\,T_0} = 529\,m.s^{-1} \qquad [2.112]$$

$$h_0 = \frac{\gamma r T_0}{\gamma - 1} = \frac{a_0^2}{\gamma - 1} = 7.10^5\ J \qquad [2.113]$$

$$h_0 = \frac{V_{max}^2}{2} \qquad [2.114]$$

$$V_{max} = 1183\ ms^{-1} \qquad [2.115]$$

The critical state corresponds to $M = 1$

In the table, we find:

$$\frac{p_C}{p_0} = 0.5283;\ p_C = 1{,}58.10^5\ Pa \tag{2.116}$$

The critical pressure is higher than atmospheric pressure. The nozzle is therefore choked.

2) In order to calculate the leakage flow rate, we shall use the formulae given by [2.47] and [2.54]:

$$q_m = \rho_0\ S_C\ V_{\max}\ Y_C \tag{2.117}$$

The section S_C "at the throat" will be the section of the aperture of the valve. When $M = 1$, the table provides $Y_C = 0{,}2588$ [2.118]

$$S_C = \pi \frac{D^2}{4} = \pi \frac{\left(2.10^{-3}\right)^2}{4} = 3{,}14.10^{-6}\ m^2 \tag{2.119}$$

$$q_m = \rho_0\ S_C\ V_{\max}\ Y_C = 3{,}09 * 3{,}14.10^{-6} * 1183 * 0{,}2588 = 2{,}97.10^{-3}\ kg.s^{-1} \tag{2.120}$$

3) Emptying the pressure cooker.

The value of the flow rate $q_m = 2{,}97.10^{-3}\ kg.s^{-1}$ found above remains valid as long as there is liquid water in the pressure cooker. In effect, temperature is constant and the liquid–vapor equilibrium will be preserved.

A proper order of magnitude of the emptying time τ_C will therefore be given by:

$$\tau_C = \frac{m}{q_m} = \frac{0{,}5}{2{,}97.10^{-3}} = 168{,}35\ s = 2{,}8\min \tag{2.121}$$

EXAMPLE 2.4 (Supersonic flow within everyone's reach).–

A bicycle pump consists of a cylindrical body with an inner diameter $D = 2cm$, in which a piston can move when pushed by the cyclist, usually by means of a connecting rod. In order to obtain a sonic flow, this pump can be added to a converging part with an exit aperture of diameter d = 100 μm.

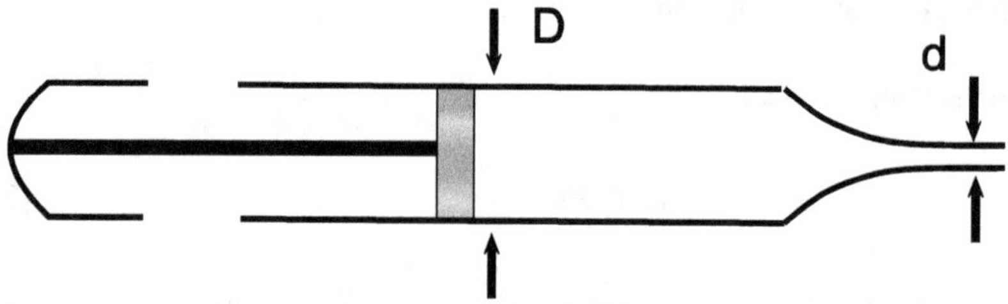

Figure 2.6. *A supersonic generator within everyone's reach*

The ambient pressure is $p_a = 1\,bar$, the temperature $T_a = 35\,°C$. The pump having already been used, the air inside is at a higher temperature $T_0 = 35\,°C$. The piston is pushed in an attempt to block the sonic flow in the aperture of diameter d.

1) As a first step, let us assume that the piston velocity is extremely low.

What is the minimum force that the cyclist is required to exert on the piston in order to have a sonic regime at the throat of diameter d?

It should be kept in mind that ambient atmospheric pressure already exerts a force on the piston.

Express this force in Newton and in kilogram-force.

One kilogram-force is, by definition, the gravitational force of 1 kg weight.

Is this force within the reach of:

– Children?

– Very well-trained athletes?

– Superman?

2) We resume the hypothesis of the very low velocity of the piston.

2.1) Given the hypotheses in question 1 (piston velocity is null), find the air mass flow rate exiting the pump.

Then, deduce the real velocity of the piston.

2.2) Assuming that the pressure inside the pump, which has been calculated to 1 is still applicable, calculate the total enthalpy of air considering first that piston velocity is negligible, then non-negligible. Deduce the validity of the hypothesis considered at 1.

Solution:

1) The outlet of the throat of the converging nozzle is at atmospheric pressure. The sonic regime is "choked" at the throat if the critical pressure is higher than or at least equal to the atmospheric pressure. The pressure of the generating state must therefore be adapted accordingly.

As usual, let us determine r:

$$r = \frac{R}{M} = \frac{8,31}{29.10^{-3}} = 285,5 \qquad [2.122]$$

In this problem, the generating state is unknown; it will be particularly determined by the force exerted on the piston. This force determines, in effect, the pressure inside the pump. Let us find the expression of the generating state in the pump. The temperature of the generating state is known $T_0 = 35°C$. The celerity of sound a_0 in the generating state and the theoretical maximum velocity in the nozzle $V_{\max}$ can therefore be deduced:

$$T_0 = 308\,K;\; a_0 = \sqrt{\gamma r T_0} = 350,8\,ms^{-1} \qquad [2.123]$$

$$V_{\max} = \sqrt{\frac{2\,a_0^2}{\gamma - 1}} = 784,4\,ms^{-1} \qquad [2.124]$$

The pressure in the pump is given by the ratio of the force exerted on the piston to its surface area. The force exerted on the piston is the resultant of the force exerted by the atmosphere and the additional force $\vec{F}$ exerted by the user:

$$p_0 = \frac{p_a S + F}{S} = p_a + \frac{F}{S};\; S = \pi\frac{D^2}{4} = 3,14.10^{-4}\,m^2 \qquad [2.125]$$

ρ_0 is deduced from the equation of state of perfect gases:

$$\rho_0 = \frac{p_0}{r T_0} \qquad [2.126]$$

The critical pressure will be at least equal to:

$$p_a = 10^5 \, Pa. \qquad [2.127]$$

The table for adiabatic flows indicates for $M = 1$ the following value for the ratio of critical pressure and pressure of the generating state:

$$\frac{p_C}{p_0} = 0,52828; \; p_0 = \frac{p_C}{0,52828} \qquad [2.128]$$

p_0 must therefore be at least equal to $p_0 = \dfrac{1}{0,52828} = 1,89 \, bar$. [2.129]

F is then deduced:

$$F = S(p_0 - p_a) = 3,14.10^{-4}(1,89.10^5 - 10^5) \qquad [2.130]$$

$$F = 27,94 \, N = 2,85 \, kgf \qquad [2.131]$$

Judging by this value, the force is within the reach of a child.

2) Considerations on the velocity of the piston

2.1) The mass flow rate at the aperture can be readily calculated, using the properties of the critical state:

$$q_m = \rho_0 V_{\max} \, s Y_C; \; Y_C = 0,2588 \qquad [2.132]$$

$$\rho_0 = \frac{p_0}{r T_0} = \frac{1,89.10^5}{285,5 * 308} = 2,15 \, kg.m^{-3} \qquad [2.133]$$

$$s = \pi \frac{d^2}{4} = 7,85.10^{-9} \, m^2; V_{\max} = 784,4 \, ms^{-1} \qquad [2.134]$$

$$q_m = \rho_0 V_{\max} \, s = 3,43.10^{-6} \, kg.s^{-1} \qquad [2.135]$$

The velocity V_P of the piston is deduced from the conservation of mass flow rate:

$$q_m = \rho_0 V_{\max} \, s Y_C = \rho_0 V_P S \qquad [2.136]$$

$$V_P = \frac{q_m}{\rho_0\ S} = 5{,}08.10^{-3}\ m.s^{-1} \qquad [2.137]$$

The total mass enthalpy of the gas in the piston is expressed by:

$$\frac{V_P^2}{2} + h_0 = \frac{V_P^2}{2} + \frac{a_0^2}{\gamma - 1} = = 2{,}58.10^{-5} + 3{,}08.10^5 = 3{,}08.10^5\ J \qquad [2.138]$$

We see that $\frac{V_P^2}{2} << \frac{a_0^2}{\gamma-1}$ [2.139]

The kinetic energy of the piston has an entirely negligible contribution to the total enthalpy. The gas can therefore be considered at rest in the body of the pump.

EXAMPLE 2.5 (Design of a hypersonic wind tunnel).–

We aim to build a hypersonic wind tunnel. The structure of this wind tunnel should be simple. A converging–diverging de Laval nozzle connects the air in the laboratory to the wind tunnel stream. The pressure of air in the laboratory is $p_0 = 1bar$ and the temperature is $T_0 = 19\,°C$. We assume that the laboratory has a very large volume and that the air is perfectly calm far enough from the inlet of the converging part. The wind tunnel stream is a sufficiently long cylindrical tube, where constant pressure $p_V < p_0$ is maintained by adequate use of a vacuum pump group.

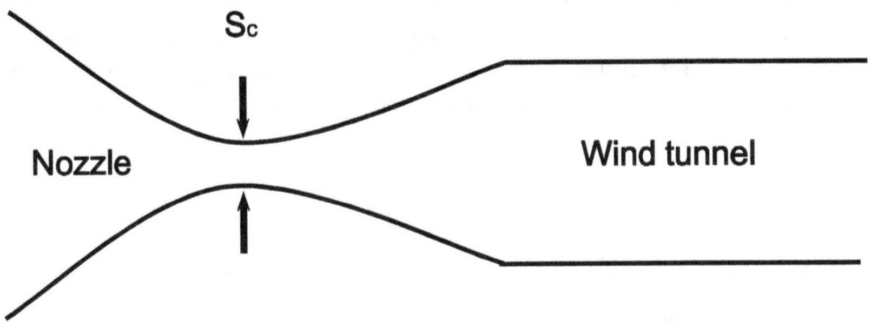

Figure 2.7. *Schematic representation of a supersonic wind tunnel*

The wind tunnel has a diameter $D = 50\ cm,$ which does not vary in this problem. Various nozzles can be adapted to it. The nozzles to be connected to the wind tunnel must all have an outlet section S_S equal to the section of the wind tunnel. The section S_C of the throat varies from one nozzle to the other.

1) A first nozzle is used and its throat has diameter $D_1 = 25\,cm$.

1.1) Find the value of r for the air. Find the celerity of sound a_0 in the laboratory.

1.2) We aim to obtain a supersonic flow in the wind tunnel. What should be the Mach number at the throat? Using it, deduce the mass flow rate of the air crossing the wind tunnel.

1.3) There should be no shock wave anywhere in the wind tunnel. What should be the pressure in the stream? Find the temperature in the stream. What is the Mach number and the air velocity in the stream?

2) We aim to have a Mach number equal to 4 in the stream. We change the nozzle.

2.1) What diameter should this new nozzle have at the throat?

2.2) What are the values of pressure, temperature and air velocity in this wind tunnel?

Solution:

Preliminary remark for better comprehension: the atmosphere of the laboratory plays the role of the generating state and the flow is determined by the conditions in the stream, where a vacuum is created. Since the stream has constant section, the velocity, temperature, pressure and density will be constant from one section to another.

1) We use a first nozzle with a throat of diameter $D_1 = 25\,cm$.

1.1) As is usual for this type of problem, several preliminary calculations are carried out as a first step.

r must be calculated.

$$r = \frac{R}{M} = \frac{8,31}{29.10^{-3}} = 285,5 \qquad [2.140]$$

Let us completely define the generating state:

$$p_0 = 1\ bar = 10^5\ Pa \quad ; \quad T_0 = 273 + 19 = 292\ K \qquad [2.141]$$

$$\rho_0 = \frac{p_0}{r T_0} = \frac{10^5}{285,5 * 292} = 1,2\, kg.m^{-3} \qquad [2.142]$$

The celerity of sound in the laboratory, which is a property of the generating state, can be calculated as:

$$a_0 = \sqrt{\gamma r T_0} = 341.6\, m.s^{-1} \qquad [2.143]$$

1.2) The flow between the generating state (null velocity) and the throat is necessarily subsonic. As we know, the transition between subsonic and supersonic flow can only take place at the throat. For the flow in the stream to be supersonic, a transition must take place at the throat. This transition involves the blockage of the sonic regime at the throat. The flow at the throat is therefore sonic (critical flow). The corresponding Mach number is equal to 1. The mass flow rate q_m at the throat of a nozzle can be calculated by the expression:

$$q_m = \rho_0\, S_C\, V_{max}\, Y_C \qquad [2.144]$$

The velocity extremum that is physically and theoretically possible (V_{max}) is equal to:

$$V_{max} = \sqrt{\frac{2\gamma}{\gamma - 1}\frac{p_0}{\rho_0}} = a_0 \sqrt{\frac{2}{\gamma - 1}} = 763,8\, m.s^{-1} \qquad [2.145]$$

The section at the throat is: $S_C = \pi \dfrac{D_1^2}{4} = 4,9.10^{-2}\ m^2$ [2.146]

The "table for adiabatic flows" indicates for $\gamma = 1.4$ on the line $M = 1$:

$$Y_C = 0,2588$$

It then follows:

$$q_m = \rho_0\, S V_{max}\, Y_C = 1,2 * 4,9.10^{-2} .* 763,8 .* 0,2588 = 11,62\, kg.s^{-1} \qquad [2.147]$$

1.3) The outlet flow at the throat must be supersonic and isentropic simultaneously. Given the conservation of flow rate, the same result must be obtained when calculated at the throat and in the outlet section of the nozzle S_S, which is the same as the section of the stream, S :

$$S = S_S = \pi \frac{D^2}{4} = 0.196\, m^2 \qquad [2.148]$$

$$q_m = \rho_0\, S\, V_{\max}\, Y_C = \rho_0\, S\, V_{\max}\, Y_S \qquad [2.149]$$

where Y_S is the value of Y at the nozzle outlet. It is worth noting that ρ_0 and $V_{\max}$ are characteristic to the generating state and they remain constant along the nozzle as long as the flow is adiabatic and reversible.

Under adiabatic conditions, the conservation of flow rate can be expressed more simply by:

$$S_C\, Y_C = S\, Y = S_S\, Y_S \qquad [2.150]$$

Let us insist on the fact that this relation is not valid for two sections separated by a shock wave. This happens essentially because the shock wave changes the generating state.

We deduce Y_S. Each value of Y_S can be found in two distinct lines of the table for adiabatic flows. This can be explained by the shape of the curve $Y\left(\frac{p}{p_0}\right)$, as mentioned in the lecture notes.

This curve has two parts: a supersonic one and a subsonic one. We aim to determine a supersonic flow after the throat of the nozzle. We must choose the value of Y corresponding to Mach number greater than 1.

Y_S is given by:

$$Y_S = \frac{S_C\, Y_C}{S_S} = 0.2588 \frac{D_1^2}{D^2} = 6,47.10^{-2} \qquad [2.151]$$

In the tables, we shall find this value on two lines:

The line where $M = 0{,}15$, $\frac{p}{p_0}$ is then equal to $0{,}98$.

The line where $M = 2{,}94$, $\frac{p}{p_0}$ is then equal to $0{,}298$.

The first line corresponds to a subsonic flow; therefore, it does not satisfy our needs. Let us stress in passing that if we choose to fix the pressure in the stream at $\frac{p}{p_0} = 0{,}98$, we shall have a subsonic flow in the diverging part of the nozzle. We have to fix the pressure p_S in the outlet section of the nozzle, hence in the stream, at:

$$p_S = 0{,}298\, p_0 = 0{,}298.10^5\ Pa = 0{,}298\, bar. \qquad [2.152]$$

This is clearly a vacuum.

2) A Mach number equal to 4 should be obtained in the stream. We change the nozzle.

2.1) In this case, the Mach number at the outlet of the nozzle is imposed. It is therefore at this location that we impose a value $Y = Y_S$.

When $M = 4$, $Y_S = 0{,}02414$, which is given in the table.

Since Y_C always has the same value at the throat, the new section S'_C of the throat will be imposed, and the diameter D_2 can be readily deduced:

$$S'_C\, Y_C = S_S\, Y_S \qquad [2.153]$$

$$S'_C = \frac{S_S\, Y_S}{Y_C} = 0.196\frac{0{,}02414}{0{,}2588} = 1{,}83.10^{-2}\ m^2 \qquad [2.154]$$

$$D_2 = \sqrt{\frac{4\, S'_C}{\pi}} = 0{,}152\, m = 15{,}2\, cm \qquad [2.155]$$

The flow rate is imposed. We shall calculate it for the outlet section:

$$q_m = \rho_0 S_S\, V_{\max}\, Y_S \qquad [2.156]$$

$$q_m = 1{,}2.0{,}196 * 763{,}8 * 0{,}02414 = 4{,}33\, kg.s^{-1} \qquad [2.157]$$

2.2) For $M = 4$, we have $\frac{p}{p_0} = 6,59.10^{-3}$ [2.158]

The pressure in the stream will therefore be $p_S = 6,59.10^{-3}\ bar$. [2.159]

The table provides the data for determining the density.

On the line $M = 4$, we find:

$$\frac{\rho}{\rho_0} = 0,02766 \tag{2.160}$$

$$\rho_S = 0,02766 * 1,2 = 3,32.10^{-2}\ kg.m^{-3} \tag{2.161}$$

The temperature is given by:

$$\frac{T}{T_0} = 0,2381 \tag{2.162}$$

$$T_S = 69,52\ K \tag{2.163}$$

The velocity V in the stream can be deduced from the direct expression of the mass flow rate:

$$q_m = \rho\, S_S\, V;\ V = \frac{q_m}{\rho S_S} = \frac{4,33}{3,32.10^{-2}.0,96} = 665,42\ m.s^{-2} \tag{2.164}$$

These results can be verified by calculating the celerity of sound in the stream:

$$a = \sqrt{\gamma r T} = 167\ ms^{-1} \tag{2.165}$$

$$V = Ma = 667\ ms^{-1} \tag{2.166}$$

This corroborates the previous calculation (with one slight difference, which can be explained by the inevitable series of roundings during calculation).

EXAMPLE 2.6 (Evaluation of missile engine performance).–

The engine to be studied is a rocket engine. It must power a missile. It is composed of:

a) An engine system that generates a gaseous flow. This engine system behaves like a reservoir containing gas at a pressure $p_{01} = 40\,bar$ and temperature $T_{01} = 3000\,K$.

This gas, whose composition is often not ecological, has a molar mass of 40 g. Its polytropic exponent is $\gamma = 1,4$.

With the exception of cryogenic engines, products such as dimethylhydrazine are often used. The value of the polytropic exponent is retained especially for the reader's convenience.

b) This engine powers a converging–diverging de Laval nozzle, with a diameter at the throat $D_C = 10\,cm$ and an outlet diameter $D_S = 40\,cm$.

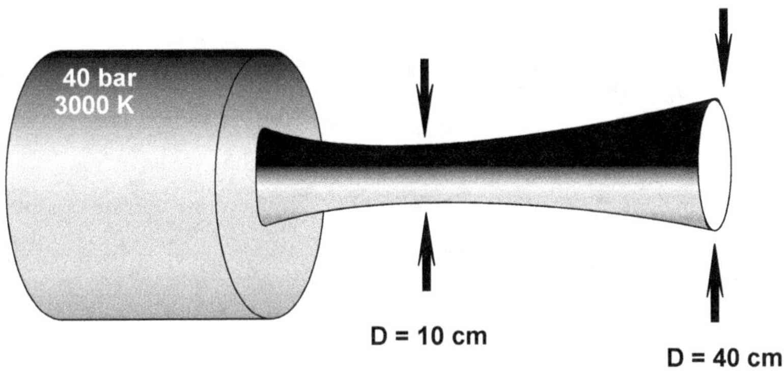

Figure 2.8. *Small missile engine*

The conditions correspond to *a sonic regime "choked" at the throat.*

1) What is the velocity of gas at the throat?

2) What is the velocity threshold that the gas cannot exceed regardless of the point of the nozzle?

3) What is the mass flow rate q_m in the nozzle?

Solution:

1) As is usual for this type of problem, we have to use the properties of the generating flow. Let us determine them at this stage:

$$r = \frac{R}{M} = \frac{8,31}{40.10^{-3}} = 207,7 \quad [2.167]$$

$$p_0 = 40\,bar = 4.10^6\ Pa; T_0 = 3000\,K \quad [2.168]$$

The gas is considered perfect:

$$\frac{p_0}{\rho_0} = r\,T_0 \quad [2.169]$$

$$\rho_0 = \frac{p_0}{r\,T_0} = \frac{4.10^6}{207,7*3000} = 6,42\,kg.m^{-3} \quad [2.170]$$

The critical state corresponds to $M = 1$. From the table, we have:

$$\frac{p_C}{p_0} = 0,5283; \frac{\rho_C}{\rho_0} = 0,6339; \frac{T_C}{T_0} = 0,8333 \quad [2.171]$$

$$p_C = 2,11.10^7\ Pa \quad [2.172]$$

$$\rho_C = 4,07\,kg.m^{-3} \quad [2.173]$$

$$T_C = 2500\,K \quad [2.174]$$

The other parameters can be deduced. The sound celerity in critical state and the critical enthalpy are:

$$a_C = \sqrt{\gamma r T_C} = 852.6\,m.s^{-1} \quad [2.175]$$

$$h_C = \frac{a_C^2}{\gamma - 1} = 1,82.10^6\ J \quad [2.176]$$

1) By definition, in the critical state, $V_C = a_C$, and therefore $M = 1$. The velocity at the throat is equal to sound celerity in the critical state, hence:

$$V_C = a_C = 852.6\, m.s^{-1} \qquad [2.177]$$

2) The maximum velocity V_{max} that the gas in a nozzle cannot physically exceed corresponds to the case in which the entire specific enthalpy h_0 of the state of the tank is transformed into specific kinetic energy:

$$a_0 = \sqrt{\gamma r T_0} = 934\, m.s^{-1} \qquad [2.178]$$

$$h_0 = \frac{a_0^2}{\gamma - 1} = 218{,}1.10^6\, J \qquad [2.179]$$

$$h_0 = \frac{V_{max}^2}{2}\,;\, V_{max} = 2088\, ms^{-1} \qquad [2.180]$$

3) In order to calculate the leakage flow rate, we shall resort to the usual formula: $q_m = \rho_0\, S_C\, V_{max}\, Y_C$

S_C is the section "at the throat".

$$S_C = \pi \frac{D^2}{4} = \pi \frac{(0{,}1)^2}{4} = 7{,}85.10^{-3}\, m^2 \qquad [2.181]$$

For $M = 1$, the table gives $Y_C = 0{,}2588$

$$q_m = \rho_0 S_C\, V_{max}\, Y_C = 6{,}42 * 7{,}85.10^{-3} * 2088 * 0{,}2588 = 27{,}23\, kg.s^{-1} \qquad [2.182]$$

EXAMPLE 2.7 (Nitrogen tank leakage).–

All the flows in this problem are considered one-dimensional, adiabatic and reversible, except for eventual singularities. Let us consider a leak-tight tank containing nitrogen at the pressure $p_0 = 3\, bar$. Nitrogen has a molar mass $M = 28\, g$. Its polytropic exponent is $\gamma = 1{,}4$.

This tank is immersed in an outer atmosphere at pressure $p_a = 1\, bar$.

The temperature is $T_0 = 20\, °C$ both inside and outside the tank.

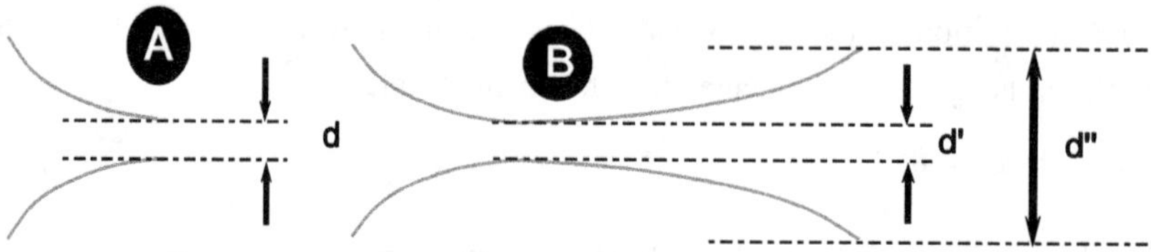

Figure 2.9. *Leakages from a nitrogen tank*

1) Completely define:

Generating state ($p_0, \rho_0, T_0, a_0, h_0$)

Corresponding critical state ($p_C, \rho_C, T_C, a_C, h_C$).

2) Being highly corroded, the tank has a hole, whose geometry is represented in the above figure (profile A). The outlet diameter of the hole opening in the air is $d = 1\,mm$.

2.1) Compare the atmospheric pressure and p_C. What conclusion can be drawn?

2.2) Calculate the nitrogen leakage mass flow rate. How many moles of nitrogen are lost per hour?

3) The hole is filled by welding, but a second hole (profile B) appears.

$d' = 1,4\,mm$ is the diameter of the hole at its minimum.

$d'' = 5,2\,d'$ at its outlet in the atmosphere.

3.1) What is the minimum pressure of the atmosphere outside the tank for which the flow remains subsonic in the hole?

3.2) What is the leakage mass flow rate with this new hole when atmospheric pressure is equal to 1 bar? How many moles of nitrogen are lost per hour in this case?

Solution:

1) Definition of the generating state.

Obviously, it is the state of the gas contained in the tank. It is a perfect gas with known pressure and temperature. The state is therefore completely defined.

Let us first determine r:

$$r = \frac{R}{M} = \frac{8,31}{28.10^{-3}} = 296,8 \qquad [2.183]$$

$$\frac{p_0}{\rho_0} = r\,T_0 \qquad [2.184]$$

$$T_0 = 20 + 273 = 293\,K \qquad [2.185]$$

$$\rho_0 = \frac{p_0}{r\,T_0} = \frac{3.10^5}{296,8.293} = 3,45\,kg.m^{-3} \qquad [2.186]$$

Starting from temperature, we know how to deduce the sound celerity (equation [2.10]) and the specific enthalpy (equation [2.14]) of this state:

$$a_0 = \sqrt{\gamma r T_0} = 348.9\,m.s^{-1} \qquad [2.187]$$

$$h_0 = \frac{a_0^2}{\gamma - 1} = 3,04.10^5\,J \qquad [2.188]$$

From these, we deduce the maximum theoretical velocity possible, which is indispensible when calculating the leakage mass flow rate.

$$h_0 = \frac{V_{max}^2}{2};\, V_{max} = 780,2\,ms^{-1} \qquad [2.189]$$

The critical state corresponds to $M = 1$. Based on the table for adiabatic flows established for $\gamma = 1.4$, we can write:

$$\frac{p_C}{p_0} = 0,5283;\, \frac{\rho_C}{\rho_0} = 0,6339;\, \frac{T_C}{T_0} = 0,8333 \qquad [2.190]$$

and the following values are obtained:

$$p_C = 1,58.10^5\,Pa \qquad [2.191]$$

$$\rho_C = 2,19\,kg.m^{-3} \qquad [2.192]$$

$$T_C = 244.2\,K \qquad [2.193]$$

The rest of the parameters are then deduced, namely the celerity of sound at the throat and the specific enthalpy of the critical state:

$$a_C = \sqrt{\gamma r T_C} = 318,54\, m.s^{-1} \tag{2.194}$$

$$h_C = \frac{a_C^2}{\gamma - 1} = 2,54.10^5\, J \tag{2.195}$$

2.1) The critical pressure is higher than the atmospheric pressure.

We have already explained in 2.2.5, 2.2.6 and in Example 2.1 that in this case the nozzle is blocked. The sonic flow is "frozen" at the throat. In other terms, the nozzle is choked.

2.2) The flow rate in a section S is expressed as:

$$q_m = \rho_0 S V_{\max} Y = \rho_0 S_C V_{\max} Y_C \tag{2.196}$$

Let us make the calculations at the throat. We find in the table that, at the throat, the critical value of Y for $M = 1$ is $Y_C = 0.2588$

The section of the throat is known: $S = \pi \dfrac{d^2}{4} = 7,85.10^{-7}\, m.s^{-1}$ [2.197]

We therefore have:

$$q_m = 3,45 * 780,2 * 7.85.10^{-7} * 0,2588 = 5,47.10^{-4}\, kg.s^{-1} = 0,55\, g.s^{-1} \tag{2.198}$$

which corresponds to $n = \dfrac{0,55}{28} * 3600 = 70,7\, mol.h^{-1}$ [2.199]

3) De Laval nozzle

3.1) The minimum sought-for pressure corresponds to an adaptation of this outlet pressure when the nozzle is choked.

The outlet section is:

$$S_S = \pi \frac{d''^2}{4} = 2,12.10^{-5}\, m^2 . \tag{2.200}$$

At the throat, the section is $S_C = \pi \dfrac{d'^2}{4} = 1,54.10^{-6}\ m^2$ [2.201]

In this case, the flow rate conservation leads to:

$$q_m = \rho_0 S V_{\max} Y = \rho_0 S_C V_{\max} Y_C = \rho_0 S_S V_{\max} Y_S \qquad [2.202]$$

or

$$SY = S_C Y_C;\ Y = \frac{S_C}{S} Y_C \qquad [2.203]$$

Therefore, Y has a value imposed at the nozzle outlet Y_S :

$$Y_C = 0,2588;\ Y_S = \frac{S_C}{S_S} Y_C; Y_S = \frac{1,54.10^{-6}}{2,12.10^{-5}} 0,2588 = 1,88.10^{-2} \qquad [2.204]$$

which corresponds to an approximate value of $M = 0,04$.

We deduce that $\dfrac{p_S}{p_0} = 0,999$ and the value of outlet pressure when the sonic flow is frozen:

$$\frac{p_S}{p_C} = 0,999 * 3.10^5 = 2,997.10^5\ Pa = 2,997\, bar \qquad [2.205]$$

We see that under real conditions, the flow cannot result as subsonic.

3.2) The critical pressure is the same as in 2.

It is therefore higher than atmospheric pressure. The nozzle is therefore choked.

There is another way to be sure in this respect: At the outlet, the ratio $\dfrac{p_a}{p_0} = \dfrac{1}{3}$. The table then gives us $Y = 0,2286$. Let us note that this value is on a line where $M = 1,43 > 1$.

Let us note that $\frac{p_a}{p_0} = \frac{1}{3} < 0,5283$, which is the value leading to the maximum value of Y. We are positively on the supersonic part of the curve $Y\left(\frac{p}{p_0}\right)$. Let us calculate the flow rate at the throat. Let us recall that the critical value of Y for $M = 1$ is $Y_C = 0,2588$:

$$q_m = \rho_0 \, S_C \, V_{\max} \, Y_C = 3,45 * 780,2 * 1,54.10^{-6} * 0,2588 = 1,07.10^{-3} \; kg.s^{-1} = 1,07 \, g.s^{-1} \qquad [2.206]$$

which corresponds to $n = \frac{1,07}{28} * 3600 = 137 \, mol.h^{-1}$ [2.207]

3

One-dimensional Compressible Flows: Irreversible Flows

3.1. Introduction

Chapter 2 was dedicated to *fully adiabatic and reversible flows*. The examples treated in this chapter indicate the limits of the universal reversibility hypothesis. In particular, the choked nozzles are subjected to the constraint of an imposed outer pressure.

This chapter focuses first on a study of *straight shock waves,* which are the prototype for the emergence of irreversibility in compressible flow.

The shock waves will then be studied from the perspective of nozzle adaptation.

3.2. Irreversible flow: straight shock wave

3.2.1. *Establishing the fundamental relations*

a) Shock wave

A shock wave is a discontinuity that divides a flow into two regions. The incoming flow is supersonic and the outgoing flow is subsonic. Physically speaking, this is a very thin area (several micrometers in thickness). It is modeled here by a plane. To set the ideas, let us consider the flow through a pipe of constant section. The shock wave is a region of the flow where phenomena can no longer be isentropic. It can be considered (which is the case in the following problems) a singularity within a flow obeying the isentropic hypotheses.

The wave propagates at supersonic velocity in a so-called "upstream" environment. Let us denote this celerity of propagation by V_1 in an environment that is assumed still.

In order to write the corresponding equations, we choose a reference system that is at rest relative to the wave.

The phenomenon is modeled by a supersonic incoming flow (on the left) of velocity V_1 and by a subsonic flow (on the right) of velocity V_2.

The wave is a discontinuity of velocity, pressure, density and temperature.

The upstream and downstream flows being considered isentropic, it will be possible to define a generating state for each of these flows.

NOTE.– This generating state is *a priori* different for upstream and downstream of the wave. This will be confirmed in the following.

Let us denote the upstream parameters by V_1, p_1, ρ_1, T_1 and the downstream parameters by V_2, p_2, ρ_2, T_2, in line with common notations.

NOTE.– At this stage, it can be anticipated that a non-isentropic process cannot preserve the generating state while the wave propagates. This will be subsequently verified.

We shall therefore distinguish the generating-state parameters of the upstream flow from those of the downstream flow by denoting them $p_{01}, \rho_{01}, T_{01}, a_{01}$ and $p_{02}, \rho_{02}, T_{02}, a_{02}$, respectively.

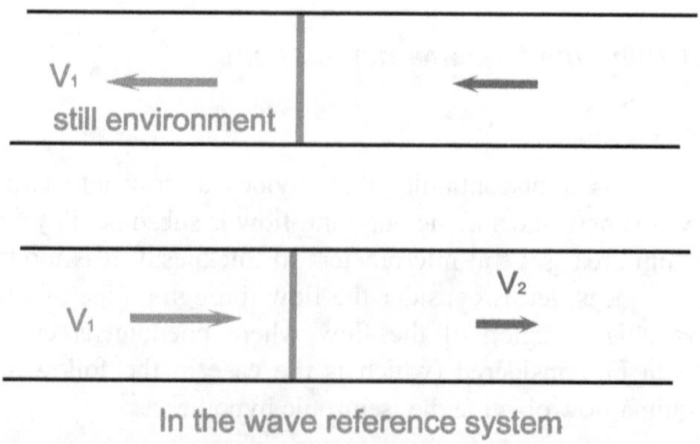

Figure 3.1. *Shock wave in a tube*

b) Writing the principles

No variation of the flow section is observed when the wave propagates. Therefore, the section S is eliminated from the continuity equation. The three principles can be written as:

Rate flow conservation

$$\rho_1 V_1 = \rho_2 V_2 \quad [3.1]$$

Equation of dynamics (according to Euler's first theorem)

$$\rho_1 V_1^2 + p_1 = \rho_2 V_2^2 + p_2 \quad [3.2]$$

Energy equation, known as Zeuner's equation

$$\frac{V_1^2}{2} + h_1 = \frac{V_2^2}{2} + h_2 \quad [3.3]$$

c) Useful relations

As a result of sometimes cumbersome manipulations of these equations, which will not be detailed here, a set of relations of practical interest are established.

A fundamental parameter is the wave Mach number $M_1 = \frac{V_1}{c_1}$ [3.4]

Prandtl established the relation:

$$V_1 V_2 = a_c^2 \quad [3.5]$$

where a_c is the critical velocity of the incoming flow.

The parameters upstream and downstream of the wave, namely pressure, density and temperature are related by the wave Mach number M_1:

$$\frac{p_2}{p_1} = 1 + \frac{2\gamma}{\gamma + 1}\left(M_1^2 - 1\right) \quad [3.6]$$

$$\frac{\rho_2}{\rho_1}=\frac{V_1^2}{a_c^2}=\frac{\frac{\gamma+1}{2}M_1^2}{1+\frac{\gamma-1}{2}M_1^2} \qquad [3.7]$$

$$\frac{T_2}{T_1}= M_1^{-2}\left[1+\frac{2\gamma}{\gamma+1}\left(M_1^2-1\right)\right]\left[1+\frac{\gamma-1}{\gamma+1}M_1^2-1\right] \qquad [3.8]$$

The Mach number behind the wave can also be determined. This number is defined starting from the fluid velocity downstream of the wave, in the wave reference system:

$$M_2^2=\frac{1+\frac{\gamma-1}{2}M_1^2}{\gamma M_1^2-\frac{\gamma-1}{2}} \qquad [3.9]$$

d) Discontinuity of the generating state

We can define two generating states, for the upstream and downstream flows, respectively.

These generating states are characterized by:

$$h_{01}=\frac{V_1^2}{2}+h_1 \qquad [3.10]$$

$$h_{02}=\frac{V_2^2}{2}+h_2 \qquad [3.11]$$

Taking into account Zeuner's equation:

$$h_{01}=h_{02}=c_p T_{01}=c_p T_{02}; T_{01}=T_{02} \qquad [3.12]$$

The shock wave preserves the temperature of the generating state.

On the contrary, the pressures and densities of the generating state vary between upstream and downstream of the wave:

$$\frac{p_{02}}{p_{01}}=\frac{\rho_{01}}{\rho_{02}}=\left[1+\frac{2\gamma}{\gamma+1}\left(M_1^2-1\right)\right]^{\frac{-1}{\gamma-1}}\left[1-\frac{2\gamma}{\gamma-1}\frac{M_1^2-1}{M_1^2}\right]^{\frac{-\gamma}{\gamma-1}} \qquad [3.13]$$

NOTE.– This last expression will prove essential for locating a shock wave in a nozzle.

These relations are used for establishing straight shock wave tables, an example of which, calculated for $\gamma = 1.4$ can be found in the appendix to this book.

3.2.2. *Applications*

EXAMPLE 3.1 (A diligent student).–

A student who is learning fluid mechanics wants to set up an experiment as an illustration of the shock waves.

He builds an air tight box fitted with a PVC tube of length $L = 3\,m$. This box and the tube are separated by a plastic diaphragm, which is designed to break when the difference between the pressures exerted on *its faces* reaches $\Delta p = 3\,bar$. The student then takes a small air compressor of the type that is normally used as tire inflator, which can be bought from car parts dealers, and connects it to the box. The other end of the tube opens in the free air in the laboratory.

The air pressure in the room where the experiment is carried on (therefore in the PVC tube before the diaphragm is broken) is $p_a = 1\,bar$ and the temperature is $T_a = 20\,°C$.

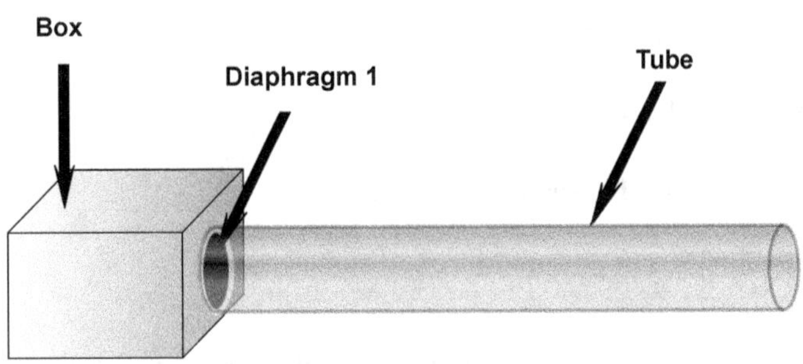

Figure 3.2. *Small shock tube*

The student then initiates the device. When the pressure in the box reaches the threshold, the diaphragm breaks and a shock wave propagates in the tube.

Air is assimilated to a gas of molar mass 29 g and polytropic exponent $\gamma = 1.4$.

1) What is the shock wave Mach number M_1? What is its celerity c_1 relative to the laboratory reference system? What is the time interval τ between the breaking of the diaphragm and the arrival of the wave at the tube outlet? Is it possible for the student to distinguish between these two phenomena?

2) What is the temperature T_2 in the tube when the wave reaches the tube outlet?

Solution:

1) The standard notations will be used, the index 1 (for undisturbed environment) being assigned to the environment upstream of the wave, and the index 2 to the environment downstream of the wave (which is affected by the wave propagation).

The table for straight shock ($\gamma = 1.4$) will be systematically used.

It is always useful to determine first the various properties of the atmosphere upstream of the wave and other parameters that may be needed when solving the problem.

The constant r is:

$$r = \frac{R}{M} = \frac{8.31}{29.10^{-3}} = 286,5 \qquad [3.14]$$

The temperature upstream of the wave is:

$$T_1 = 20C = 293\,K \qquad [3.15]$$

The celerity of sound upstream of the wave is:

$$a_1 = \sqrt{\gamma r T_1} = 342,8\,ms^{-1} \qquad [3.16]$$

With pressure p_1 being 1 bar, and given that the diaphragm breaks for a pressure difference $\Delta p = 3\,bar$, the pressure p_2 is:

$$p_2 = p_1 + \Delta p = 4\,bar = 4.10^5\ Pa \qquad [3.17]$$

at the breaking point. p_2 is also the pressure downstream of the wave.

The table for straight shock then gives for $\frac{p_2}{p_1} = 4$ a Mach number for the environment $M_1 = 1,89$.

Therefore, the celerity of the wave is:

$$c_1 = M_1 \, a_1 = 647,9 \, ms^{-1} \qquad [3.18]$$

The time needed for the wave to transit the tube of length $L = 3m$ is:

$$t_T = \frac{L}{c_1} = 4,6.10^{-3} \, s = 4,3 \, ms \qquad [3.19]$$

It will be difficult for the student to perceive this time interval, as he/she will simultaneously perceive the breaking of the diaphragm and the wave diffraction at the tube outlet.

2) The table for straight shock provides the ratio of the temperatures on the two sides of the wave.

$$\frac{T_2}{T_1} = 1,6 \, . \qquad [3.20]$$

The temperature of the gas at the tube outlet is:

$$T_2 = 1,6.293 = 469 \, K = 196 \, C \qquad [3.21]$$

EXAMPLE 3.2 (A physicist visits Nosferatu).–

A Transylvanian fluid mechanics scientist sets himself the task to destroy a vampire who has turned his neighborhood into a bloodbath.

NOTE.– According to Wikipedia, *in the Western world, a vampire is an undead being whose survival depends on drinking blood.* It can only be destroyed by silver bullets.

The vampire is a local feature in Transylvania, a region in the center of Romania, bordered by the Carpathian Mountains.

Nosferatu is a vampire character in a famous movie directed by Friedrich Murnau (1922), and remade by Werner Herzog in 1979, featuring Klaus Kinski as the leading actor.

A taste of the environment: It is winter time in the mountains, it is very cold, the temperature being 4°C below zero. The ambient pressure is 76 cmHg. Silver melts at $T_{Fus} = 961{,}93\,°C$. The density of mercury is $\rho_{HG} = 13600\,kg.m^{-3}$

The air is assimilated to a gas of molar mass $M = 29\,g$ whose polytropic exponent is $\gamma = 1{,}4$.

Understandably, the physicist is somewhat anxious, as he wonders whether velocity would cause the bullet to melt. To find the answer, he relies on the hypothesis, which will be resumed here, that a straight shock wave attaches in front of the bullet immediately as it exits the gun.

NOTE.– One can readily understand the physicist's anxiety: the risk is significant, given that a human being bitten by a vampire turns into a vampire too.

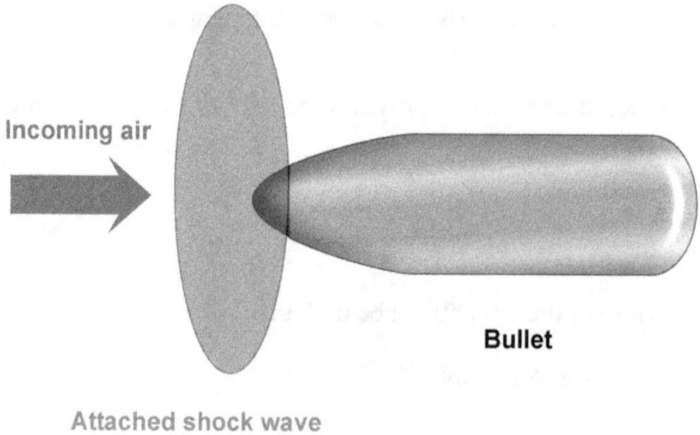

Figure 3.3. *Shock wave attached to a supersonic projectile*

1) The reference system chosen is connected to the bullet. What should be the shock wave Mach number for the bullet to melt?

2) What should then be the velocity of the bullet relative to the ground?

3) Several values for military or civilian projectiles are provided below.

In your opinion, is there a risk that the physicist's bullet melts before hitting the vampire?

Should we distrust all physicists returning from a vacation in Transylvania?

What are the projectiles on the list that should not be melted in silver?

Note.– “Die dicke Bertha” (Big Bertha), also named “Die fleissige Bertha” (“Hard-working Bertha”) was a gun used by the German army during the First World War. Having a caliber of 420 mm, it fired projectiles of 1150 kg, each with an explosive load of 144 kg, at 12,5 km. Contrary to a commonly held idea, this gun was not used in the bombardment of Paris.

Type of gun	Velocity of the projectile in m s^{-1}
KM 45 automatic gun	270
Pump-action shotgun cal. 12	381
Big Bertha	400
500 Magnum Smith and Wesson	549
Machine-gun M 249	914

Table 3.1. *Velocities of some military or civilian projectiles*

Solution:

1) The standard notations will be used, the index 1 (for undisturbed environment) being assigned to the environment downstream of the wave, and the index 2 to the environment upstream of the wave (environment affected by the wave propagation).

The table for straight shock $(\gamma = 1,4)$ will be systematically used.

When solving this type of problem, it is always useful to first calculate various properties of the atmosphere upstream of the wave and other useful parameters:

$$T_1 = -4C = 269\,K$$

$$p_1 = 0,76 * 9,81 * 13600 = 1,013.10^5\ Pa \qquad [3.22]$$

$$r = \frac{R}{M} = \frac{8.31}{29.10^{-3}} = 286,5 \qquad [3.23]$$

$$a_1 = \sqrt{\gamma r T_1} = 328,5\,ms^{-1} \qquad [3.24]$$

The silver contained by the bullet melts provided the temperature upstream of the wave is $T_2 = T_{Fus} = 961,93C = 1234,9\,K$ [3.25]

In this case, the ratio of upstream temperature to downstream temperature of the wave is:

$$\frac{T_2}{T_1} = \frac{1234,9}{269} = 4,59 \qquad [3.26]$$

The straight shock table indicates for this ratio a Mach number $M_1 = 4,34$

2) This yields a celerity of the wave that is equal to the velocity of the bullet relative to the atmosphere:

$$c_1 = M_1\, a_1 = 328,5 * 4,34 = 1426\, ms^{-1} \qquad [3.27]$$

3) It can be seen that the velocity required for the bullet to melt is considerably below that of the projectiles fired by a wide variety of guns.

We can therefore be reassured as to the harmlessness of all physicists returning from Transylvania, provided they carry a gun!

EXAMPLE 3.3 (A hunting session).–

Throughout the present problem, we shall assume, as a rough simplification, that when a sphere travels at supersonic velocity through a gas, the attached shock wave is straight and perpendicular to the path of the given sphere.

A hunter chases a hare through a forest in Northern Europe. It is cold; the temperature is $T_a = 4\,°C$. Before leaving, the hunter checked the barometer, which indicated $p = 74,9\, cm\ Hg$.

The hunter makes his own gunpowder and ammunition. The ammunition consists of small spheres, usually made from lead, which are fired at a velocity $V_1 = 820\, m.s^{-1}$.

1) Show that the hunting bullets are supersonic projectiles

Calculate the temperature at which the shock wave carries the hunting bullet. What is the pressure around this bullet?

2) Ammunition change

One day, being short of lead, the hunter uses tin balls as ammunition. Though a skilled hunter who is generally successful, when using tin bullets he returns empty-handed. Explain why.

Numerical data

Properties of the air:

Under a pressure $p_a = 1 bar$ and $T_a a = 0°C$, the molar mass is $M = 29 g$

Polytropic exponent $\gamma = 1,4r$

Properties of various metals

Lead melting temperature: $T_{FusPb} = 326 °C$

Tin melting temperature: $T_{FusSn} = 232 °C$

Solution:

1) Let us determine the various properties of the atmosphere upstream of the wave.

$$r = \frac{R}{M} = \frac{8,31}{29.10^{-3}} = 286,5 \quad [3.28]$$

$$T_1 = 4C = 277 K \quad [3.29]$$

$$p_1 = 0,749 * 9,81 * 13600 = 0,999.10^5 \ Pa \quad [3.30]$$

The celerity of sound in the atmosphere in which the ball moves:

$$a_1 = \sqrt{\gamma r T_1} = 332,8 ms^{-1} \quad [3.31]$$

The celerity of the wave c_1 is given by the velocity of air relative to the ball to which it is attached. Therefore, it is equal to the velocity of the ball:

$$c_1 = 820 ms^{-1} \quad [3.32]$$

The wave Mach number will therefore be: $M_1 = \frac{c_1}{a_1} = 2,46$ [3.33]

The ball is a supersonic projectile and the shock wave hypothesis is therefore validated.

The upstream to downstream ratios of temperatures and pressures will therefore be provided by the straight shock wave table. The upstream temperatures and pressures are those around the ball and can be deduced:

$$\frac{T_2}{T_1} = 2{,}09; \frac{p_2}{p_1} = 6{,}89 \quad [3.34]$$

$$T_2 = 2{,}09 * 277 = 579\,K = 306\,C \quad [3.35]$$

$$p_2 = 6{,}89 * 0{,}999.10^5 = 6{,}88.10^5\,Pa \approx 6{,}9\,bar \quad [3.36]$$

2) The temperature $T_2 = 306\,C$ of the gas surrounding the ball is below the lead melting temperature, but higher than the tin melting temperature.

We can therefore expect the hunter to fire hunting bullets, as these retain their genuine hardness. This way he will reach his target. On the contrary, tin ammunition melts and misses the target due to its softness. Consequently, the hunter returns with no game.

EXAMPLE 3.4 (A factory tragedy).–

In a chemical plant, two halls H_1 and H_2 are connected by a passage of length $L_1 = 40\,m$. The dimensions that are useful in this problem are indicated in the figure.

The hall H_1 is separated from the passage by a light partition A. The passage opens directly into the hall H_2. A glass window B is facing this entry at a distance $L_2 = 20\,m$.

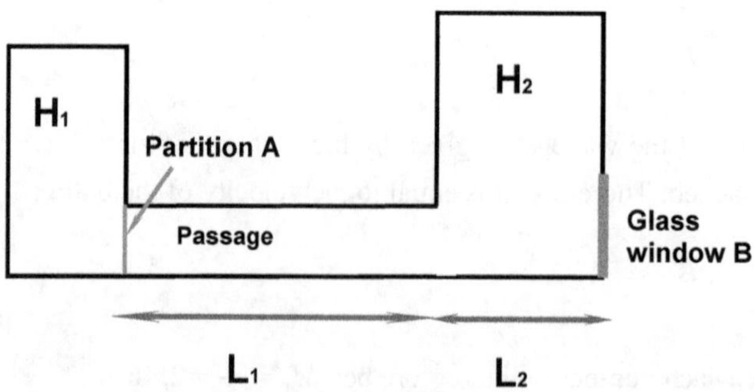

Figure 3.4. *Industrial accident*

Under normal conditions, the pressure is $p_a = 1\,bar$ and the temperature is $T_a = 25°C$ in H_1, H_2 and the passage.

Because of a fire, the pressure and temperature in H_1 increase.

The walls of H_1 resist, but there is a weak point in the structure, namely the partition A.

When the pressure reaches $p_1 = 3\,bar$ in H_1, the partition A blows up, and a shock wave propagates through the passage, and then in H_2. The rise in pressure up to 3 bar and the related increase in temperature is assumed to be a transformation at constant volume (an isochoric process).

It can be assumed that the values of temperature and pressure in H_2 and in the passage in front of the wave are equal to the initial values. Moreover, the distance L_2 being relatively short, we shall neglect the effects of shock wave diffraction and shall consider that it propagates in H_2 the same way as in the passage.

NOTE.– A change of section generates a distortion of the shock wave, which is no longer plane, nor one dimensional. The approximation made here is acceptable if we intend to find a time order of magnitude.

1) What is the shock wave Mach number M_1 in the passage?

How long after the destruction of A will it take before the glass window B is also destroyed?

2) What temperature T_2 will be registered in the passage behind the wave?

Solution:

1) Let us first calculate various properties of the atmosphere upstream of the wave. The wave obviously propagates from H_1 to H_2. The initially still fluid in the passage is therefore in front of the wave. The variables that are considered relative to it are assigned index 1. The explosion determines the values of the fluid behind (downstream of) the wave, which are assigned index 2.

As a consequence:

$$T_1 = 25C = 298\,K \qquad [3.37]$$

$$p_1 = 1\,bar = 10^5\ Pa \qquad [3.38]$$

We calculate r as:

$$r = \frac{R}{M} = \frac{8,31}{29,10^{-3}} = 286,5 \qquad [3.39]$$

and the celerity of sound in the undisturbed atmosphere:

$$a_1 = \sqrt{\gamma r T_1} = 345,7\,ms^{-1} \qquad [3.40]$$

The pressure downstream of the wave resulting from the explosion:

$$p_2 = 3\,bar = 3.10^5\ Pa \qquad [3.41]$$

The ratio $\frac{p_2}{p_1} = 3$ allows us to determine the wave Mach number M_1, by means of the straight shock wave table.

We find $M_1 = 1,65$ [3.42]

This Mach number is by definition the ratio of the wave celerity in the downstream environment, consisting of the passage and H_2, namely c_1, and the celerity of sound in this environment, namely a_1.

The wave celerity is then calculated: $c_1 = M_1\,a_1 = 570,4\,ms^{-1}$ [3.43]

The wave reaches the glass window with a celerity c_1, travelling a distance $L = L_1 + L_2 = 40 + 20 = 60\,m$ in a time interval $t_B = \frac{L}{c_1} = 0,105\,s$ [3.44]

2) The straight shock wave table also indicates on the line $M_1 = 1,65$ the ratio $\frac{T_2}{T_1} = 1,42$. [3.45]

Therefore, the temperature of the gas behind the wave is:

$$T_2 = 1,42 * 298 = 423,2\,K = 150,2\,C \qquad [3.46]$$

NOTE.– This explains why this type of explosion is destructive, the heat effects and the mechanical effects of overpressure being combined.

EXAMPLE 3.5 (Measurement of thermocouple response time).–

The response time of small thermocouples is sometimes measured in laboratories with the device described here.

The response time is defined as the time needed by the thermocouple to reach the temperature of the surrounding air.

NOTE.– A thermocouple is composed of two different metals welded together. The welding generates a voltage that is first-order proportional to the temperature. This voltage is low (the Chromel–Alumel thermocouple, which is one of the most sensitive, generates a signal of approximately 40 μV per degree). One of the common usages of the thermocouple is in "local instantaneous" measurements. This is, for example, the case of combustion flows, which are predominantly turbulent. The measurement is often complicated by radiation problems that will not be treated here. In order to have a local measurement and reduce the response time, the thermocouple must be small. The response time is never null. Nevertheless, it may be small, and this justifies the research methods on response time, similar to the one presented here.

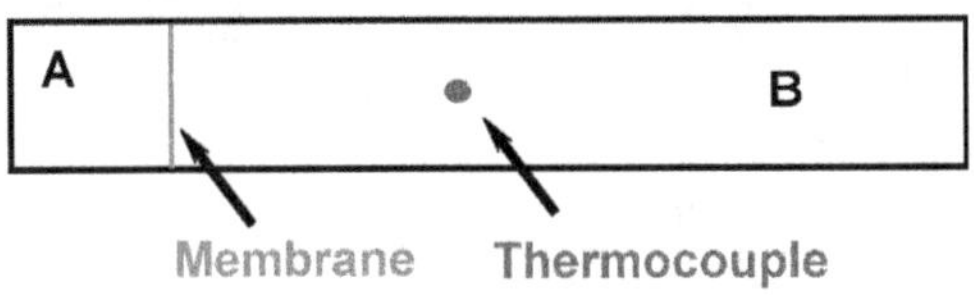

Figure 3.5. *Thermocouple testing device*

A tube consists of two chambers A and B separated by a membrane. The chamber B is at atmospheric pressure P_a = $1\,bar$ and ambient temperature $T_a = 20\,°C$.

The pressure in chamber A increases and the membrane breaks very abruptly when the pressure in A reaches $p_2 = 10\,bar$.

The thermocouple to be tested is set in chamber B. Upon the passage of the shock wave generated by the membrane breaking, the temperature of the air surrounding the thermocouple sharply increases. We shall attempt to evaluate this increase in temperature and the time needed for this increase, as registered by the thermocouple.

1) Calculate:

The Mach number, M_1

The celerity of the shock wave, c_1

The temperature T_2 that is consequently reached by the gas in the chamber.

2) The thermocouple to be tested has the shape of a sphere of diameter $d = 50\,\mu m$. We shall evaluate the time of the temperature scale imposed on the thermocouple, which will be assimilated to the time of contact between the surface of the discontinuity represented by the shock wave, and the thermocouple.

Find the value τ of this time duration.

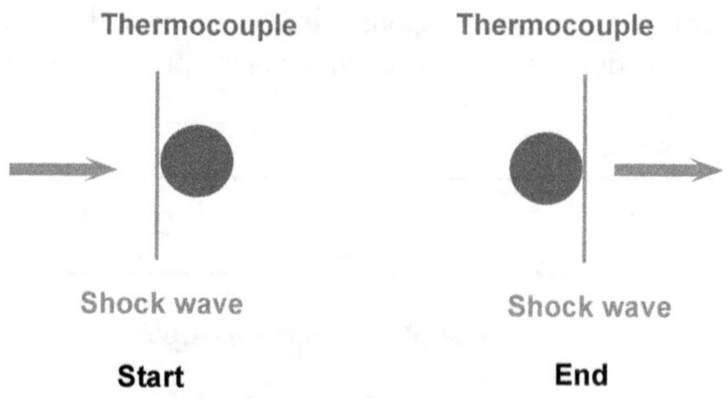

Figure 3.6. *Measurement of the response time of a thermocouple*

The thermal response time of a thermocouple of this type is of the order of tens of milliseconds. Do you think that the method employing a shock wave as a means to increase the air temperature is appropriate when measuring the response time of such thermocouples?

Solution:

1) Let us first determine the various properties of the atmosphere upstream of the wave.

The wave propagates from A to B. The gas parameters in B before the passage of the wave are therefore assigned index 1. The parameters applied to the thermocouple downstream of the wave are assigned index 2.

Let us calculate r:

$$r = \frac{R}{M} = \frac{8,31}{29.10^{-3}} = 286,5 \qquad [3.47]$$

Upstream of the wave (where the gas is "fresh"):

$$T_1 = 20C = 293K;\ p_1 = 1\,bar = 10^5\ Pa \qquad [3.48]$$

The celerity of sound upstream of the wave is:

$$a_1 = \sqrt{\gamma r T_1} = 342,8\,ms^{-1} \qquad [3.49]$$

The pressure downstream of the wave is $p_2 = 10\,bar = 10^6\ Pa$

The ratio $\frac{p_2}{p_1} = 6$ allows us to determine the wave Mach number M_1, by means of the straight shock wave table. The result is $M_1 = 2,3$ [3.50]

This Mach number is, by definition, the ratio of the wave celerity in chamber B, c_1, to the celerity of sound in this environment, a_1. Therefore, the celerity is $c_1 = M_1\,a_1 = 788,4\,ms^{-1}$ [3.51]

From the straight shock wave table, we also obtain on the line for $M_1 = 2,3$ the ratio $\frac{T_2}{T_1} = 1,95$

The temperature of the gas behind the wave is:

$$T_2 = 1,95 * 293 = 571,3\,K = 298,3\,C \qquad [3.52]$$

2) The wave propagates through the sphere at a velocity equal to $c_1 = 788,4\,ms^{-1}$ along its diameter $d = 50.10^{-6}\,m$

in a time $\tau_t = \dfrac{d}{c_1} = 6,34.10^{-8}\,s$ [3.53]

The temperature of the wall of the thermocouple rises to $298,3\,°C$ within a duration of the order of 600 μs, compared to the response time of thermocouples, which is of the order of milliseconds. The heating is considered instantaneous relative to the response time related to the propagation of the thermal wave inside the metal.

NOTE.– The response time of the thermocouple can be obtained by registering the voltage response. This will be considered as a response at pure temperature step.

EXAMPLE 3.6 (Quick chemistry of high temperatures).–

In order to obtain a rapid chemical reaction, a hydrocarbon, assumed to be a perfect gas of molar mass 100 g, is placed in a shock tube of constant section and diameter $D = 10\,cm$. Before the passage of the shock wave, the gas (considered "fresh gas") temperature is $T_1 = 300\,K$. The gas polytropic exponent is $\gamma = 1,4$. The wave is supposed to increase the gas temperature to $T_2 = 1200\,K$.

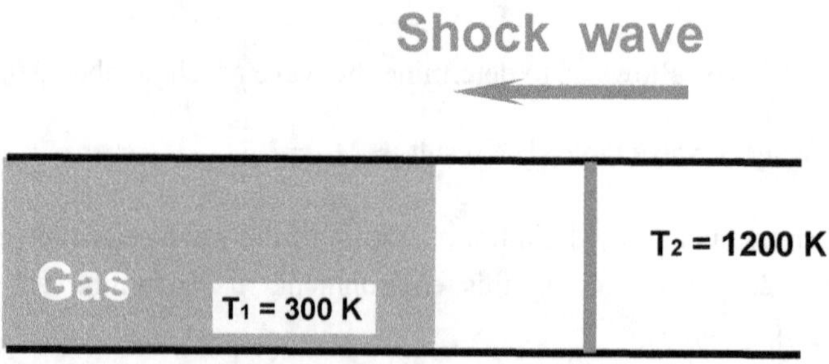

Figure 3.7. *High-temperature chemistry*

1) What values should the Mach number M_1 and the celerity c_1 of this wave in the fresh gases have to render this rise in temperature possible?

2) What volume of gas is it possible to heat up to $T_2 = 1200\,K$ within < 1 ms?

Solution:

1) Let us determine the useful properties of the atmosphere upstream of the wave

$$r = \frac{R}{M} = \frac{8,31}{100.10^{-3}} = 83,1 \qquad [3.54]$$

$$T_1 = 300\,K \,;\, a_1 = \sqrt{\gamma r T_1} = 186,8\,ms^{-1} \qquad [3.55]$$

The pressures (upstream and downstream) are not relevant to this problem.

The ratio of upstream to downstream temperature:

$$\frac{T_2}{T_1} = \frac{1200}{300} = 4 \qquad [3.56]$$

leads to calculating, by means of the straight shock waves table, the required wave Mach number:

$$M_1 = 3,97 \qquad [3.57]$$

The celerity of the wave is $c_1 = M_1 a_1 = 741,6\,ms^{-1}$ [3.58]

2) The volume V_{OL} whose temperature can be increased to $T_2 = 1200\,K$ is the volume swept by the wave during this time. It is determined by the area of the tube section $S = \pi\dfrac{(0.1)^2}{4} = 7,85.10^{-3}\,m^2$

[3.59]

and the distance that the wave propagates in 1 ms.

$$V_{OL} = 7,85.10^{-3} * 741,6 * 10^{-3} = 5,82.10^{-3}\,m^3 = 5,82\,l \qquad [3.60]$$

From a chemical point of view, it is clear that the shock tube facilitates the transfer of substantial enthalpy to a non-negligible volume of reactive gas within a very short time. To reach this performance, the chemist has to set up an unusually complex experimental device.

3.3. Partially irreversible flows: shock wave in a nozzle

3.3.1. *Change of the generating state by the shock wave*

In 2.2, we treated nozzles throughout which the flow was reversible. In this case, when the nozzle is choked, the value of the function $Y\left(\frac{p}{p_0}\right)$ at the outlet $Y_S\left(\frac{p_S}{p_0}\right)$ is fixed. The outlet pressure p_S is therefore imposed in order to satisfy this value of Y_S. This cannot be achieved in practice.

NOTE.– In the case of an engine outlet nozzle, given that the outlet pressure is the atmospheric pressure, this would require setting a flight altitude, and even weather conditions!

Given that p_S is imposed only by means of its ratio to the pressure p_0 of the generating state, a different method can be applied.

Let us denote the pressure of the generating state determining the flow rate at the throat by p_{01} (p_{01} was previously denoted by p_0 when there was only one possible generating state).

The value of Y_S now fixed for the "choked" rate flow imposes a value of $\frac{p_S}{p_{01}}$, which cannot be achieved by p_S, which is physically dictated by the outlet conditions.

Is it then possible to imagine that a new value of the pressure of the generating state (p_{02}) can be imposed at the nozzle outlet so that the value of $Y_S = Y\left(\frac{p_S}{p_{02}}\right)$ can be attained?

This question *cannot be answered in fully* reversible *flow*.

On the contrary, it is known that a *shock wave modifies the generating state*. A shock wave that is correctly positioned in the diverging part of the nozzle can adapt the outlet pressure to the imposed pressure. At the same time, it will constitute a passage in the diverging part of the nozzle, between an upstream supersonic flow and a downstream subsonic flow.

In fact, this wave is automatically created and naturally positioned.

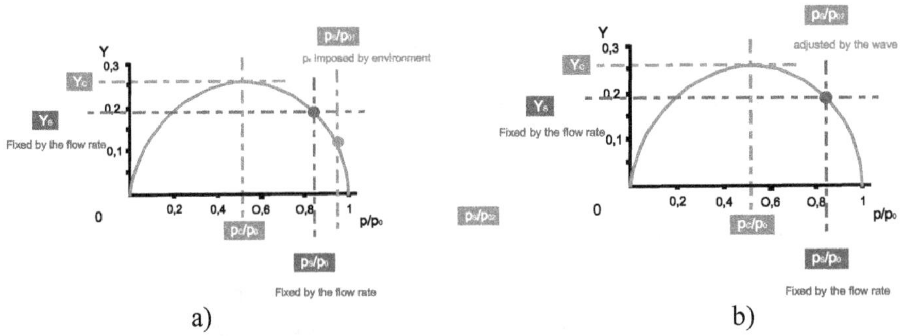

Figure 3.8. *Outlet pressure adjustment by the shock wave. a) Upstream of the shock wave; b) Downstream of the shock wave*

How is this position determined?

The method is simple.

The generating state is given, and its pressure is p_{01}.

The nozzle is choked and the flow rate is imposed:

$$q_m = \rho_0 V_{\max} S_C Y_C \qquad [3.61]$$

The value of Y at the outlet Y_S is determined by:

$$Y_S = Y_C \frac{S_C}{S_S} \qquad [3.62]$$

which imposes the ratio $\frac{p_S}{p_{02}}$. The pressure p_S is fixed, the pressure of the new generating state p_{02} is imposed and can be determined starting from the values of the function Y.

For this purpose, we can use the tables for adiabatic flows, along the columns $\frac{p}{p_0}$ and Y.

The ratio $\frac{p_{01}}{p_{02}}$ is then known. The straight shock wave table allows us to calculate the shock wave Mach number M_1.

The shock wave is positioned in the diverging part of the nozzle, where the flow has this value for the Mach number. The table for adiabatic flows again proves useful: it provides the values for the Mach number and the sections ratio $\frac{S}{S_C}$.

It is obvious that *the previous method is only applicable to choked nozzles,* in which the flow in the diverging part is supersonic.

In the case of a subsonic flow, there can be no shock wave: the flow upstream of a shock wave is necessarily supersonic. A subsonic flow is possible only for one downstream pressure, as stated in 2.1.

3.3.2. *Applications*

EXAMPLE 3.7 (Irreversible De Laval nozzle).–

A very large reservoir contains air at a pressure of 15 bar and a temperature of 300 K. This reservoir is connected to another reservoir by a converging–diverging nozzle. The pressure p_s in this second reservoir is adjustable.

The nozzle has a section $S_C = 2\,cm^2$ at the throat. Its outlet section S_S is equal to $4\,S_C$ C.

The molar mass of the air is $M = 29\,g$. Its polytropic exponent is $\gamma = 1{,}4$.

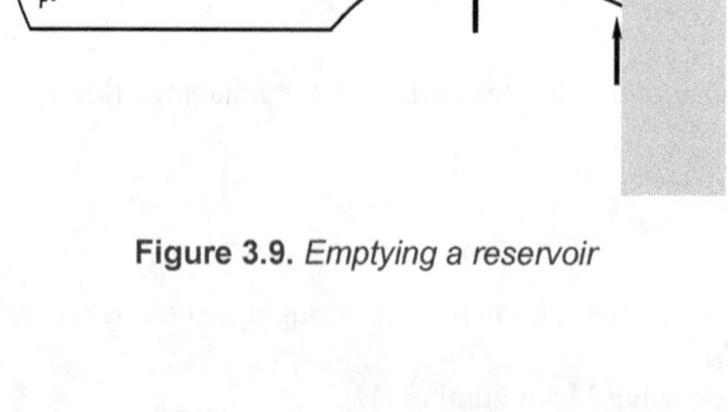

Figure 3.9. *Emptying a reservoir*

1) Calculate the various characteristics of the critical state corresponding to this generating state: pressure p_0, temperature T_0 and density ρ_0. Calculate the celerity of sound a_0 in the generating state (first reservoir), the maximum velocity V_{max} that the gas can reach in the nozzle.

2) When the sonic regime is reached, what will the velocity of the gas be for Mach 1 $(M = 1)$? Where can it be measured?

3) Calculate the mass flow rate in the nozzle when the sonic flow is choked.

4) We want to adjust the pressure p_S in the second reservoir in order to obtain a subsonic flow throughout the nozzle. What is the permitted range of values for p_S?

5) We want to have a supersonic regime at the outlet of the nozzle without creating shock waves in the nozzle or outside of it. What pressure p_S should be fixed?

6) We fix a new pressure p_S (that we will seek to determine in this question) above the pressure found in (5), but outside the range found in (4). We then observe a shock wave that is stabilized in the nozzle in a location where the section is $S_{Shock} = 2\,S_C$.

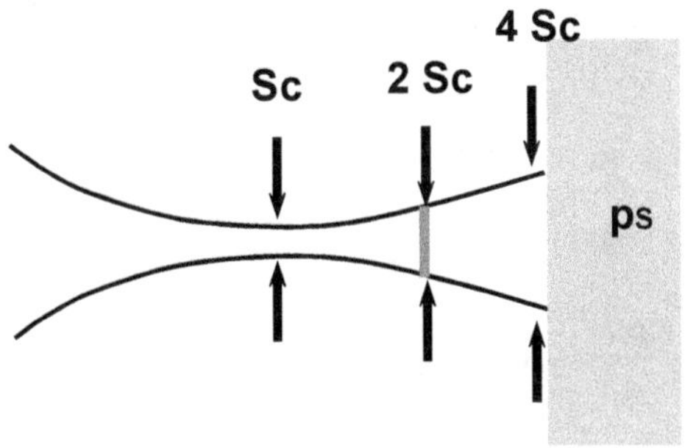

Figure 3.10. *Adjustment of the position of the shock wave in de Laval nozzle*

6.1) What is the Mach number of the air arriving in the shock wave?

6.2) Find the characteristics (pressure, temperature and density) of the new generating state downstream of the shock wave.

6.3) What is the Mach number downstream of the shock wave in its vicinity?

6.4) What is the pressure p_S corresponding to these flow conditions?

Solution:

1) This question relates to determining the generating state, which comes first when solving a nozzle problem.

The existence of a reservoir facilitates the calculation:

$$r = \frac{R}{M} = \frac{8,31}{29.10^{-3}} = 286,5\,SI \quad [3.63]$$

$$p_0 = 15\,bar = 1,5.10^6\,Pa : T_0 = 300\,K \quad [3.64]$$

$$\frac{p_0}{\rho_0} = r\,T_0 \quad [3.65]$$

$$\rho_0 = \frac{p_0}{r\,T_0} = \frac{1,5.10^6}{286,5 * 300} = 17,45\,kg.m^{-3} \quad [3.66]$$

$$a_0 = \sqrt{\gamma r\,T_0} = 346,9\,ms^{-1} \quad [3.67]$$

$$V_{\max} = a_0\sqrt{\frac{2}{\gamma - 1}} = 775,7\,ms^{-1} \quad [3.68]$$

2) The critical state corresponds to $M = 1$

From the table, we have:

$$\frac{p_C}{p_0} = 0,5283; \frac{\rho_C}{\rho_0} = 0,6339; \frac{T_C}{T_0} = 0,8333 \quad [3.69]$$

$$p_C = 7,92.10^5\,Pa \quad [3.70]$$

$$\rho_C = 11,06\,kg.m^{-3} \quad [3.71]$$

$$T_C = 249,9\,K \quad [3.72]$$

The celerity of sound can thus be deduced, and it coincides with the fluid velocity V_C for $M = 1$

$$a_C = V_C = \sqrt{\gamma r T_C} = a_0\sqrt{\frac{2}{\gamma + 1}} = 316,6\,m.s^{-1} \quad [3.73]$$

According to the Hugoniot theorem, this velocity can only be reached at the throat.

3) The mass flow rate through each section can be calculated from:

$$q_m = \rho_0 \; V_{\max} \; SY \tag{3.74}$$

It is logical to calculate it at the throat, where we know the section $S_C = 2.10^{-4}\ m^2$ and

$$Y_C = 0,2588 \text{ for } \gamma = 1,4 \tag{3.75}$$

$$q_m = \rho_0 \; S_C \; V_{\max} \; Y_C \;= 17,45 * 775,7 * 2.10^{-4} * 0,2588 = 0,7\,kg.s^{-1} \tag{3.76}$$

4) If a reversible flow is to be obtained in the downstream reservoir in subsonic regime, this is possible for all the pressures p_S ranging from $p_S = p_0$, which is the flow threshold at zero flow rate, and the pressure corresponding to choking at the sonic throat. The whole range of these downstream pressures covers the subsonic part of the $Y = Y\dfrac{p}{p_0}$ curve. When the sonic is choked, the flow rate is then fixed, and Y takes an imposed value $Y = Y_S\left(\dfrac{p_S}{p_0}\right)$ in the outlet section.

The flow rate conservation leads to:

$$q_m = \rho_0 \; V_{\max} \; S_C \, Y_C = \rho_0 \; V_{\max} \; S_S \, Y_S ; \; S_C \, Y_C = \; S_S \, Y_S ; \; Y_S = 6,47.10^{-2} \tag{3.77}$$

The second limit of pressure will be such that $Y = Y_S\left(\dfrac{p_S}{p_0}\right) = Y_S = 6,47.10^{-2}$ at the outlet. Following a simple linear interpolation, from the table we have

$$\frac{p_S}{p_0} = 0,9872, \text{ which leads to } p_S = 1,481.10^6 \; Pa = 14,8\,bar \tag{3.78}$$

In subsonic regime, the flow exits the nozzle reversibly for all the pressures ranging between $p_S = 14,8\,bar$ and $p_S = 15\,bar$.

NOTE.– This last value constitutes a theoretical "threshold", since it corresponds to a null flow rate!

5) For a reversible exit in supersonic regime, a similar reasoning is employed. This time, only one representative point will be found on the supersonic part of the $Y = Y_S\left(\frac{p_S}{p_0}\right)$ curve. In effect, in order to have a supersonic regime in the diverging region, the sonic regime must be choked at the throat and the mass flow rate is imposed. Therefore, only one value of Y_S is adequate.

For the same value $Y_S = 6.47.10^{-2}$ and a Mach number between 3,01 and 3,02, we find after interpolation

$$\frac{p_S}{p_0} = 2,662.10^{-2} \qquad [3.79]$$

and hence $p_S = 0,4\,bar$. [3.80]

6) For any pressure ranging between $p_S = 0,4\,bar$ and $p_S = 14,8\,bar$, the regime is sonic at the throat and the mass flow rate is still imposed at q_m = $0,7\,kg.s^{-1}$. By conserving the same generating state, the value of $Y_S = 6,47.10^{-2}$ is still imposed.

For a p_S value that differs from the previous two values, we should still have $\frac{p_S}{p_0} = 2,662.10^{-2}$, which is impossible for a fully reversible flow.

Therefore, a phenomenon modifying p_0 should occur in the diverging region of the nozzle.

This will be a shock wave.

In this problem, we fix the position of the shock wave in a point where the section $S_{Shock} = 2S_C$.

6.1) For $\frac{S}{S_0} = 2$, the table indicates a Mach number

$$M = 2,2. \qquad [3.81]$$

6.2) The ratio of generating-state upstream pressure p_0 to downstream pressure p_{02} of the shock wave is $\frac{p_0}{p_{02}} = 0,6281$. [3.82]

Therefore, $p_{02} = 2,39.10^6\ Pa$. [3.83]

Given that the energy is conserved during wave propagation (the shock phenomenon is irreversible, but non-dissipative), it is known that:

$$T_{02} = T_0 \quad [3.84]$$

$$a_{02} = a_0 \quad [3.85]$$

$$V_{\max 2} = V_{\max 1} \quad [3.86]$$

Similarly, we find

$$\rho_{02} = \frac{p_{02}}{r\,T_{02}} = 27,8\,kg.m^{-3} \quad [3.87]$$

A new outlet value for Y_S is then obtained as:

$$q_m = \rho_{02}\ V_{\max 2}\ S_C\ Y_C \quad [3.88]$$

$$Y_S = 4,058.10^{-2} \quad [3.89]$$

The table then indicates:

$$\frac{p_S}{p_{02}} = 0,01428;\ p_S = 3,413.10^4\ Pa = 0,34\,bar \quad [3.90]$$

6.3) For $M = 2,2$, the table indicates a downstream Mach number:

$$M_2 = 0,5471 \quad [3.91]$$

It is thus verified that the flow downstream of the wave is definitely subsonic.

EXAMPLE 3.8 (Small propeller study).–

In the previous examples, the flow emerged from an identifiable reservoir. This was a deliberate choice aimed at facilitating the resolution of problems that were focused on specific issues.

In many practical cases, the flow in a nozzle is generated by a device with complex geometry. To allow for a more simple resolution of the problems posed by this type of gas generator, we consider a "fictitious state", from which the flow emerges through a reversible adiabatic process. The method for determining this generating state starting from a flow known for a given section is exemplified below.

A small propeller prototype is composed of a combustion chamber with an arbitrary structure, which generates a flow of combustion products in a tube T of diameter $d = 1\,cm$. The molar mass of the combustion products is $M_m = 70$. For this gas, the polytropic exponent is 1,4. In the tube, their pressure is $p_1 = 9\,bar$ and their temperature is $T_1 = 1200\,C$. The flow rate of these gases is $q_m = 1500\,kg.h^{-1}$.

A device designed to fly in an atmosphere of pressure $p_a = 20\,000\,Pa$ is fitted with this engine.

Between the chamber and the tube T, there is a converging nozzle with a throat of diameter d'. This nozzle is choked, and it controls the flow rate (the sonic regime is "choked at the throat").

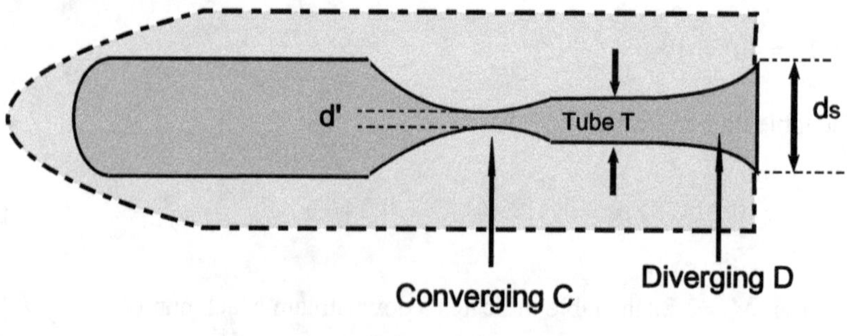

Figure 3.11. *Diagram of the propeller*

1) Calculate the velocity of the combustion products in T, and then their Mach number.

2) Deduce from the above the generating state of the system whose pressure, density and temperature will be denoted by p_0, ρ_0 and T_0, respectively.

3) What is the diameter d' of the nozzle throat? What is the velocity V_C of the gases at this throat?

4) A diverging part that opens to the outside through a section S_S of diameter d_S is added to tube T. What should d_S be so that the nozzle outlet pressure is equal to the outside pressure? What engine thrust corresponds to this configuration?

Solution:

1) In this example, the flow is generated by a complex system that does not originate in a reservoir. The generating flow is "fictitious" and can be deduced from the flow data known for a section through its pressure, temperature and rate flow.

Let us first determine the density ρ_1 and velocity V_1 of the flow.

For this perfect gas, we have:

$$r = \frac{R}{M_m} = \frac{8,31}{70.10^{-3}} = 118,7\, SI \qquad [3.92]$$

$$\frac{p_1}{\rho_1} = r\, T_1 \qquad [2.5]$$

$$p_1 = 9.10^5\; Pa;\; T_1 = 1200 + 273 = 1473\, K \qquad [3.93]$$

$$\rho_1 = \frac{9.10^5}{118,7 * 1473} = 5,15\, kg.m^{-3} \qquad [3.94]$$

Let us write the expression of the mass flow rate:

$$q_m = \rho_1\, S_1\, V_1 \qquad [3.95]$$

$$q_m = \frac{1500}{3600} = 0,417\, kg.s^{-1};\; S_1 = \pi\frac{10^{-4}}{4} = 7,85.10^{-5}\; m^2 \qquad [3.96]$$

$$V_1 = \frac{0,417}{5,15 * 7,85.10^{-5}} = 1031,5\, m.s^{-1} \qquad [3.97]$$

The celerity of sound in the tube T is:

$$a_1 = \sqrt{\gamma r T_1} = \sqrt{1,4 * 118,71 * 1473} = 494,7\, m.s^{-1} \qquad [3.98]$$

The Mach number M_1 in the tube is:

$$M_1 = \frac{V_1}{a_1} = 2,08 \qquad [3.99]$$

2) We know the state of the gas in the tube:

$$p_1 = 9.10^5\, Pa;\ \rho_1 = 5,15\, kg.m^{-3};\ T_1 = 1473\, K \qquad [3.100]$$

On the line corresponding to Mach number $M_1 = 2,08$, the tables for adiabatic flow indicate the relation between this state and the generating state:

$$\frac{p_1}{p_0} = 0,1128;\ \frac{\rho_1}{\rho_0} = 0,2104;\ \frac{T_1}{T_0} = 0,5361 \qquad [3.101]$$

$$p_0 = 7,98.10^6\, Pa = 79,8\, bar \qquad [3.102]$$

$$\rho_0 = 24,48\, kg.m^{-3} \qquad [3.103]$$

$$T_0 = 2748\, K \qquad [3.104]$$

It is useful to also calculate the celerity of sound in the generating state:

$$a_0 = \sqrt{\gamma r T_0} = 675,8\, m.s^{-1} \qquad [3.105]$$

3) The adiabatic flows table indicates the ratio of the section S_1 for which the Mach number is $M_1 = 2,08$ to the critical section S_C of the same pipe where the sonic regime is established:

On the line $M_1 = 2,08$, we can observe:

$$\frac{S}{S_C} = 1,8056 \qquad [3.106]$$

We shall have the sonic regime for the section $S_C = \dfrac{S_1}{1,8056}$ [3.107] which determines the diameter of the nozzle throat:

$$d'^2 = \frac{d^2}{1,8056}; d' = 7,44.10^{-3}\, m = 0,744\, cm \quad [3.108]$$

We verify that this diameter is smaller than d.

The critical state corresponds to the sonic regime. Here the velocity of the gases is equal to the local celerity of sound

$$V_C = a_C \quad [3.109]$$

This celerity of sound at the throat can be calculated starting from the temperature at the throat. The ratio of this critical temperature to that of the generating state is provided by the table, on the line corresponding to the sonic regime, $M = 1$.

It indicates: $\dfrac{T_C}{T_0} = 0,8333$ [3.110]

This yields:

$$T_C = 0,8333 * 2748 = 2290\, K \quad [3.111]$$

$$a_C = \sqrt{\gamma r T_C} = 616,9\, m.s^{-1} \quad [3.112]$$

Therefore, the local velocity of the gases at the throat is:

$$V_C = 616,9\, m.s^{-1} \quad [3.113]$$

4) The outlet pressure p_S has to be equal to the atmospheric pressure $p_a = 20\,000\, Pa$. If the opposite were true, the phenomenon would be irreversible (straight or oblique shock wave, expansion wave, etc.), adapting the outlet flow to the surrounding atmosphere.

S_S can be readily calculated by the direct use of tables. In effect, in the outlet section, we can calculate:

$$\frac{p_S}{p_0} = \frac{2.10^4}{7.89.10^6} = 2,53.10^{-3} \quad [3.114]$$

A rapid interpolation yields:

$$M_S = 4,753 \quad [3.115]$$

A further interpolation allows us to find the ratio of the outlet section to the critical section (of diameter d'):

$$\frac{S_S}{S_C} = 20,53 \quad [3.116]$$

$$\frac{d_S}{d'} = \sqrt{20,53} = 4,53;\ d_S = 4,53\, cm \quad [3.117]$$

The same table also indicates the ratio of the outlet local velocity to the critical velocity:

$$\frac{V_S}{V_C} = 2,216 \quad [3.118]$$

The outlet velocity of gases is:

$$V_S = 2,216 * 616,9 = 1367,1\, m.s^{-1} \quad [3.119]$$

The engine thrust P can be readily deduced (see equation [7.11]):

$$P = q_m V_S = 0,417 * 1367,1 = 570,1\, N = 58,1\, kgf \quad [3.120]$$

This value of the thrust may seem low. It is worth noting that the engine operates at high altitude, therefore in low-pressure atmosphere, where aerodynamic resistance is low.

3.4. Conclusion

It is worth noting here that the study of one-dimensional compressible flows constitutes a very simplifying approach to both flows and shock waves. The slightest pipe singularity (notably an abrupt change of section) poses serious difficulties to these hypotheses. Similarly, the adiabatic character of the pipes can be applied to a ceramic nozzle, but less so for a metallic tube. The outlet nozzle of a rocket engine such as Vulcain, which propels the European rocket Ariane, can be taken as an example. The diverging region is composed of hosepipes through which hydrogen flows, thereby cooling the nozzle walls by heat convection. The hypothesis referring to the adiabatic (and similarly to the one-dimensional case) character is obviously not applicable. Finally, if the friction induced in these generally high-velocity flows were to be taken into account, the reversibility hypothesis would be hardly acceptable.

The perspective of refined rocket engine design does not belong in an initiation book. As already mentioned, it is our intention to guide the reader through the resolution of simple problems, while allowing them to have a physical grasp of the principles underlying the reasoning. Moreover, we have tried to provide the reader with an elementary dimensioning tool, which is in most cases sufficient.

4

Modeling and Numerical Simulations

4.1. Introduction

Computational Fluid Dynamics (CFD) is applied in many areas of industrial interest, especially in aeronautics, and in all the fields of fluid dynamics, such as the problems of fluid–structure coupling and the flow around the wind turbine rotors.

The CFD approach presents a reliable universal method for modeling unsteady turbulence. It supplements the prototype testing and replaces the wind tunnel testing of reduced models, providing relevant data. Moreover, compared to experimental devices, it contributes to saving considerable time and costs.

For example, the data commonly obtained when applying CFD to wind turbines are the force coefficients, the limits of flow lines on the surfaces of the blades and the distributions of velocity and pressure around the rotor and near the blade. This can facilitate largely the comprehension of the aerodynamic phenomena analyzed.

The fluid flows are governed by partial differential equations. Applying CFD allows an approximation of these partial differential equations by algebraic equations, which can be solved numerically.

The CFD software that we have opted for is ANSYS Fluent. The design domain will be represented with Ansys 15.0 meshing software, and Fluent 15.0 solver will be used for numerical simulations. It is based on the finite-volume theory approach to solving the Navier–Stokes equations.

Numerical simulation is implemented in several stages, which will be detailed in this chapter.

The results of the numerical simulation are then compared with the experimental data presented in the literature, which have resulted from experiments conducted under conditions that are similar to those of the simulations in order to validate the CFD model.

4.2. Methodology description and simulation approach

The fluid flows are governed by partial differential equations representing the laws of conservation of mass, momentum and energy. In fluid mechanics, the flow regime is characterized by the Reynolds number. When the latter is small, the flow is laminar, while for Reynolds numbers higher than the order of 10^5 or 10^6, the flow becomes turbulent. The numerical resolution of the unsteady Navier–Stokes equations in turbulent regime is extremely time-consuming. Alternatively, the averaged version of Navier–Stokes equations, usually named RANS (*Reynolds-Averaged Navier–Stokes*) is used. These equations allow the effective modeling of turbulent flows while significantly reducing the numerical calculations. These turbulent flows are characterized by fields of fluctuating velocities. According to Reynolds, each unsteady variable in the Navier–Stokes equations can be decomposed into a time-averaged quantity $\bar{\phi}$ and a fluctuating quantity $\phi'(t)$:

$$\phi(t) = \bar{\phi} + \phi'(t) \qquad [4.1]$$

$$\bar{\phi} = \frac{1}{\Delta t}\int_0^{\Delta t} \phi(t)\, dt \qquad [4.2]$$

Given the low Mach number identified in the operating conditions of wind turbines, for example, the flow can be considered incompressible. The RANS equations for an incompressible fluid are reduced to:

$$\begin{cases} \nabla.\vec{U} = 0 \\ \rho \dfrac{D\vec{U}}{Dt} = -\nabla P + \mu\nabla^2\vec{U} - \rho\dfrac{\partial}{\partial x_j}\left(\overline{u_i' u_j'}\right) + \rho\vec{g} + \vec{F} \end{cases} \qquad [4.3]$$

$$\frac{D}{Dt} = \frac{\partial}{\partial t} + \left(\vec{U}.\nabla\right) \qquad [4.4]$$

where $\vec{U}$ is the velocity vector, ρ is the fluid density, P is the pressure, $\vec{F}$ is the resultant of specific forces exerted in the fluid and μ is the dynamic viscosity.

Nevertheless, the previous equations contain the turbulent correlations $\overline{u_i' u_j'}$ which are defined by Reynolds stresses and represent the effects of turbulence. They introduce additional unknowns, which are modeled for the closure of the equations for turbulent flow by means of URANS.

In FLUENT, there is a wide range of turbulence models, which are usually classified depending on the number of additional transport equations required for the calculation of turbulent viscosity. These models of turbulences are based on the hypothesis of Boussinesq [SCH 07] relating Reynolds stresses to the mean flow strain rate:

$$\tau_{ij} = -\rho \overline{u_i' u_j'} = \mu_t \left(\frac{\partial U_i}{\partial x_j} + \frac{\partial U_j}{\partial x_i} \right) - \frac{2}{3} \rho k \delta_{ij} \qquad [4.5]$$

where τ_{ij} is the Reynolds tensor, k is the turbulent kinetic energy, μ_t is the turbulent viscosity and δ_{ij} is the Kronecker delta.

Equation [4.5] highlights the presence of two variables μ_t and k that model the Reynolds stresses. In our case, a k-omega SST (Shear Stress Transport) turbulence model will be used to solve the system. This model was developed by Menter [MEN 94] who combined two *k-epsilon* and *k-omega* models so that the latter could be used in an area close to the wall and the former in the areas distant from the wall (at the center). Thanks to these characteristics, a turbulence model that is more robust and precise than the standard averaged Reynolds models (URANS) is obtained for a wide category of flows. It is generally retained as the most appropriate solution for the aerodynamic study of wind turbines and it is often recommended in the literature related to this type of computation [NOB 14, NOB 11, GAS 14, OEBO 13]. A citable example is the comparative study of various turbulence models (Standard k–ε, RNG k–ε, Realizable k–ε, Standard k–ω and the k–ω SST) elaborated by Gasser E. Hassan *et al.* [NOB 11]. They have shown that among all these turbulence models, the k–ω SST model shows the best performance and its satisfactory results are closest to the experimental results.

The k–ω SST turbulence model allows the solving of two transport equations, one for the kinetic energy of the turbulence k and the other for the specific dissipation rate of turbulent kinetic energy ω [FLU 06]:

$$\frac{\partial(\rho k)}{\partial t} + \nabla.\left(\rho k \vec{U}\right) = \nabla.(\Gamma_k \nabla k) + \tilde{G}_k - Y_k \qquad [4.6]$$

$$\frac{\partial(\rho \omega)}{\partial t} + \nabla.\left(\rho \omega \vec{U}\right) = \nabla.(\Gamma_\omega \nabla \omega) + G_\omega - Y_\omega + D_\omega \qquad [4.7]$$

where:

$\tilde{G}_k$ turbulence generated by the mean velocity gradient,

G_ω turbulence generated by ω,

Γ_k diffusivity of k,

Γ_ω diffusivity of ω,

Yk dissipation of k due to turbulence,

$Y\omega$ dissipation of ω due to turbulence,

D_ω Transverse diffusion.

The turbulent viscosity coefficient for this model is defined by:

$$\mu_t = \frac{\rho k}{\omega} \frac{1}{\max\left[\frac{1}{\alpha^*}, \frac{SF_2}{\alpha_1 \omega}\right]} \quad [4.8]$$

with:

$$\left\{\begin{array}{ll} S \equiv \sqrt{2\Omega_{ij}\Omega_{ij}}, & \Omega_{ij} = \frac{1}{2}\left(\frac{\partial U_i}{\partial x_j} - \frac{\partial U_j}{\partial x_i}\right) \\ F_2 = \tanh\left(\phi_2^{\,4}\right), & \phi_2 = \max\left[2\frac{\sqrt{k}}{0.09\omega y}, \frac{500\mu}{\rho y^2 \omega}\right] \\ \alpha^* = \alpha_\infty^*\left(\frac{\alpha_0^* + \mathrm{Re}_t/\mathrm{Re}_k}{1 + \mathrm{Re}_t/\mathrm{Re}_k}\right), & \mathrm{Re}_t = \frac{\rho k}{\mu\omega} \end{array}\right\} \quad [4.9]$$

where y is the distance perpendicular to the surface. For further details, please see ANSYS Inc. [ANS 09]. The computation results significantly depend on the value of empirical constants that are used by default in the Fluent code:

$$a_1 = 0.31,\ \mathrm{Re}_k = 6,\ \alpha^*_0 = 0.024,\ \alpha^*_\infty = 1$$

A CFD code consists of three main structures:

– Preprocessor (ANSYS)

It is used for the presentation of the design domain (creation of the geometry), the mesh generation and the definition of the type of boundary conditions.

– Processor (FLUENT Solver)

After exporting the problem into Fluent, this solver helps to numerically define the fluid operating conditions, to specify the boundary conditions and choose the iterative computation process by proposing several numerical methods (space-time discretization, numerical resolution methods, pressure–velocity coupling). This code also offers an interface for controlling the progress of computations and the rate of convergence at any moment.

– Post-processor

It helps visualize the results: (fields of velocity, pressure, turbulence and the flow lines or the paths of particles). It also allows the tracing of space-time evolution of all the variables of the flow.

The results of the simulations can be exported to ANSYS CFD-Post for more in-depth analyses.

A detailed description of all these stages of the resolutions is provided in the following section.

4.3. Modeling and simulation of coupled systems

The equations describing the fluid–structure coupled system are presented in this chapter under variational form. They are discretized by means of the finite element method, in order to obtain the matrix systems that will subsequently be numerically solved.

4.3.1. *Mathematical formulation. Behavior equations*

The equations that govern the behavior of the structure and fluid can be written as follows:

The equation of motion for the structure:

$$\nabla.\boldsymbol{\sigma} - \rho_s \ddot{\mathbf{u}} = \mathbf{0} \qquad [4.10]$$

The boundary conditions associated with the structure are:

$$\mathbf{u}\big|_{\Gamma_u} = \bar{\mathbf{u}}, \qquad \boldsymbol{\sigma}.\mathbf{n}\big|_{\Gamma_f} = \bar{\mathbf{f}} \tag{4.11}$$

The wave equation for the fluid:

$$\Delta\mathbf{P} - \frac{1}{c^2}\ddot{\mathbf{P}} = 0 \tag{4.12}$$

The boundary conditions associated with the fluid:

$$\mathbf{P}\big|_{\Gamma_P} = \bar{\mathbf{P}}, \qquad \left.\frac{\partial \mathbf{P}}{\partial \mathbf{n}}\right|_{\Gamma_a} = -\rho_f \bar{\mathbf{a}} \tag{4.13}$$

The boundary condition on the free surface Γ_{sl} is:

$$\left.\frac{\partial \mathbf{P}}{\partial \mathbf{n}}\right|_{\Gamma_{sl}} = -\frac{1}{g}\ddot{\mathbf{P}} \tag{4.14}$$

For the fluid–structure coupled systems in which the structure is fully immersed, equation [4.13] is approximated by:

$$\mathbf{P}\big|_{\Gamma_{sl}} = 0 \tag{4.15}$$

4.3.2. *Fluid–structure coupling conditions*

The boundary conditions on the fluid–structure interface denoted by C allow considering the effect of the fluid on the structure and *vice versa*. On the one hand, these are slip conditions that reflect the fact that the fluid does not penetrate the walls. Then, the velocity normal to the surface of the fluid must be equal to the velocity normal to the structure in any point of C. Expressed in terms of pressure, this equality can be written as:

$$\left.\frac{\partial \mathbf{P}}{\partial \mathbf{n}}\right|_{C} = -\rho_f \ddot{\mathbf{u}} \tag{4.16}$$

On the other hand, the continuity of the normal component of the stresses tensor should be verified for any point of C:

$$\boldsymbol{\sigma}.\mathbf{n}\big|_C = \mathbf{P}.\mathbf{n} \qquad [4.17]$$

4.4. Variational formulation

Let $\mathbf{u}^*$ be a virtual field of displacement associated to the structure such that $\forall \mathbf{u}^* / \left\{ \mathbf{u}^*\big|_{\Gamma_u} = \mathbf{0} \right\}$, and $\mathbf{P}^*$ a virtual field of pressure associated to the fluid such that $\forall \mathbf{P}^* / \left\{ \mathbf{P}^*\big|_{\Gamma_P} = \mathbf{0} \right\}$.

The integration by parts of the equations considering the boundary conditions and the coupling conditions results in the following two integral functions:

For the structure

$$\begin{aligned} \int_{\Omega_s} \rho_s \mathbf{u}^* \ddot{\mathbf{u}} \mathrm{d}V + \int_{\Omega_s} \boldsymbol{\varepsilon}^* \boldsymbol{\sigma}.\mathrm{d}V &= \int_{\Omega_s} \rho_s \mathbf{u}^* \ddot{\mathbf{u}} \mathrm{d}V + \int_{\Omega_s} \boldsymbol{\varepsilon}^* \mathbf{D} \boldsymbol{\varepsilon} \mathrm{d}V \\ &= \int_{\Gamma_f} \mathbf{u}^* \overline{\mathbf{f}} \mathrm{d}S + \int_C \mathbf{u}^* \mathbf{n} \mathbf{P} \mathrm{d}S \end{aligned} \qquad [4.18]$$

For the fluid

$$\int_{\Omega_f} \frac{1}{c^2} \mathbf{P}^* \ddot{\mathbf{P}} \mathrm{d}V + \int_{\Omega_f} \nabla \mathbf{P}^* \nabla \mathbf{P} \mathrm{d}V = \int_{\Gamma_a} \mathbf{P}^* \overline{\mathbf{a}} \mathrm{d}S - \rho_f \int_C \mathbf{P}^* \mathbf{n} \ddot{\mathbf{u}} \mathrm{d}S \qquad [4.19]$$

4.5. Finite element approximation

The fluid domain and the structure domain are subdivided into finite elements. For each element, the physical unknowns are approximated by polynomial functions. The integral functions are then discretized using the Galerkin method and integrated by each element using the Gauss method.

4.5.1. *Approximation of physical unknowns*

For an element of the structure

$$\{\mathbf{u}\} = [\mathbf{N}_s]\{\mathbf{u}\}_e \tag{4.20}$$

For an element of the fluid

$$\{\mathbf{P}\} = [\mathbf{N}_f]\{\mathbf{P}\}_e \tag{4.21}$$

4.5.2. *Integration of variational forms*

The system equations are integrated for each (structure and fluid) element considering the approximations. This leads to the following matrix expressions:

$$[\mathbf{M}]_e\{\ddot{\mathbf{u}}\}_e + [\mathbf{K}]_e\{\mathbf{u}\}_e = \{\mathbf{f}\}_e + [\mathbf{L}]_e\{\mathbf{P}\}_e \tag{4.22}$$

$$[\mathbf{E}]_e\{\ddot{\mathbf{P}}\}_e + [\mathbf{H}]_e\{\mathbf{P}\}_e = \{\mathbf{a}\}_e - \rho_f[\mathbf{L}]_e^t\{\ddot{\mathbf{u}}\}_e \tag{4.23}$$

The matrix $[\mathbf{L}]_e$ is obtained by the following integration:

$$[\mathbf{L}]_e = \int_{C_e}[\mathbf{N}_s]^t\{\mathbf{n}\}[\mathbf{N}_f]\,\mathrm{d}S \tag{4.24}$$

$[\mathbf{L}]_e$ is the elemental matrix of the fluid–structure interaction. It is the centerpiece of this modeling, since it is thanks to it that the fluid–structure coupling is numerically considered.

The following stage is marked by the assembly of elemental matrices (of the fluid domain and of the structure), which results in the following matrix expressions:

$$[\mathbf{M}]\{\ddot{\mathbf{u}}\} + [\mathbf{K}]\{\mathbf{u}\} = \{\mathbf{f}\} + [\mathbf{L}]\{\mathbf{P}\} \tag{4.25}$$

$$[\mathbf{E}]\{\ddot{\mathbf{P}}\} + [\mathbf{H}]\{\mathbf{P}\} = \{\mathbf{a}\} - \rho_f[\mathbf{L}]^t\{\ddot{\mathbf{u}}\} \tag{4.26}$$

4.6. The vibro-acoustic problem

In the compressible fluid hypothesis, the equation governing the dynamic behavior of the fluid is the wave equation, associated to the boundary and coupling conditions. The discretization and finite element approximation lead to a matrix equation [4.17] representing the acoustic behavior of the fluid, which is coupled with the structure through the boundary conditions imposed to the interface.

If the structure is also modeled by the finite element method, its dynamic behavior in the absence of dampening is represented by the expression [4.16]. In this case, the coupling appears under the effect of stresses exerted by the field of fluid pressure on the flexible walls of the structure.

Using [4.16] and [4.17], the *equations of the free coupled system* can be written as follows:

$$\begin{bmatrix} \mathbf{M} & \mathbf{0} \\ \rho_f \mathbf{L}^t & \mathbf{E} \end{bmatrix} \begin{Bmatrix} \ddot{\mathbf{u}} \\ \ddot{\mathbf{P}} \end{Bmatrix} + \begin{bmatrix} \mathbf{K} & -\mathbf{L} \\ \mathbf{0} & \mathbf{H} \end{bmatrix} \begin{Bmatrix} \mathbf{u} \\ \mathbf{P} \end{Bmatrix} = \begin{Bmatrix} \mathbf{0} \\ \mathbf{0} \end{Bmatrix} \quad [4.27]$$

The system [4.18] is not symmetrical. The modal computation of the response function is based on the matrix inversion of this system. The symmetrization is however possible by processes such as the one based on the rigidity decoupling proposed by Iron [BEN 08]. Applying this process to [4.18] yields:

$$\begin{bmatrix} \mathbf{M}+\rho_f \mathbf{L}\mathbf{H}^{-1}\mathbf{L}^t & \mathbf{L}\mathbf{H}^{-1}\mathbf{E} \\ \mathbf{L}\mathbf{H}^{-1}\mathbf{E} & \mathbf{E}^t\mathbf{H}^{-1}\mathbf{E}/\rho_f \end{bmatrix} \begin{Bmatrix} \ddot{\mathbf{u}} \\ \ddot{\mathbf{P}} \end{Bmatrix} + \begin{bmatrix} \mathbf{K} & \mathbf{0} \\ \mathbf{0} & \mathbf{E}^t/\rho_f \end{bmatrix} \begin{Bmatrix} \mathbf{u} \\ \mathbf{P} \end{Bmatrix} = \begin{Bmatrix} \mathbf{0} \\ \mathbf{0} \end{Bmatrix} \quad [4.28]$$

Let us note that if the numerical resolution of the fluid–structure coupling is conducted on MATLAB®, there is no need to apply the symmetrization process on the matrix expression [4.18], as MATLAB® automatically performs/executes this operation for small systems.

4.7. The hydro-elastic problem

If the fluid is considered incompressible, the terms proportional to $1/c^2$ are annulled, as the celerity is very high $(c \to \infty)$ in this case, and the equation [4.16] becomes:

$$[\mathbf{H}]\{\mathbf{P}\} = -\rho_f [\mathbf{L}]^t \{\ddot{\mathbf{u}}\} \quad [4.29]$$

In the case of free vibrations, the equation of the coupled problem can be expressed only as a function of the degrees of freedom of the structure, which results in the following expression:

$$([\mathbf{M}]+[\mathbf{Ma}])\{\ddot{\mathbf{u}}\}+[\mathbf{K}]\{\mathbf{u}\}=\{\mathbf{0}\} \quad [4.30]$$

with

$$[\mathbf{Ma}]=\rho_f[\mathbf{L}][\mathbf{H}]^{-1}[\mathbf{L}]^t \quad [4.31]$$

where $[\mathbf{Ma}]$ is the matrix called "added mass matrix".

4.8. Applications

EXAMPLE 4.1 (flow in a pipe).–

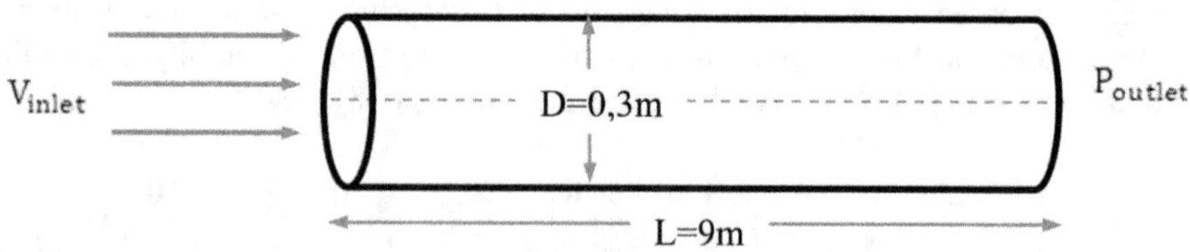

Figure 4.1. *Flow in a pipe*

Let us consider a fluid flowing through a pipe of circular cross-section of constant radius. The diameter of the pipe is D = 0,3 m and its length is L = 9 m. The wind velocity is imposed at the domain inlet (V_{inl} = 2 m.s^{-1}). The atmospheric pressure is imposed at the outlet (P_{outlet} = 1 atm). The Reynolds number depends on the diameter of the computational domain (D), the dynamic viscosity (μ), the fluid density (ρ) and the flow velocity V_{inlet}:

$$Re=\frac{\rho V_{inl} D}{\mu} \quad [4.32]$$

Depending on the value of Reynolds number, there are two main regimes: a laminar flow regime (Re < 1000) and a turbulent regime (Re > 10000).

The objective is to solve this problem numerically using ANSYS FLUENT to determine:

– the velocity distribution in the pipe,

– the pressure contour,

– the velocity profile at the pipe outlet,

– the pressure axial distribution.

Solution:

For the sake of simplicity, we shall treat an axisymmetric problem. Once the problem is solved for half of the pipe, the results can be displayed for the whole pipe.

In a first stage, let us consider a laminar flow in a pipe. The density is $\rho = 1{,}225$ kg.m^{-3} (Re = 375). The computational domain and the boundary conditions are presented in Figure 4.2.

– *Geometrical model setup*

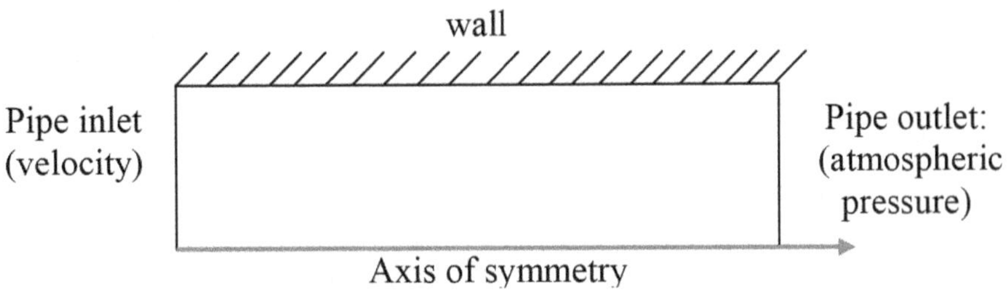

Figure 4.2. *Computational domain*

– *Mesh*

We have used a structured mesh, given its geometrical simplicity

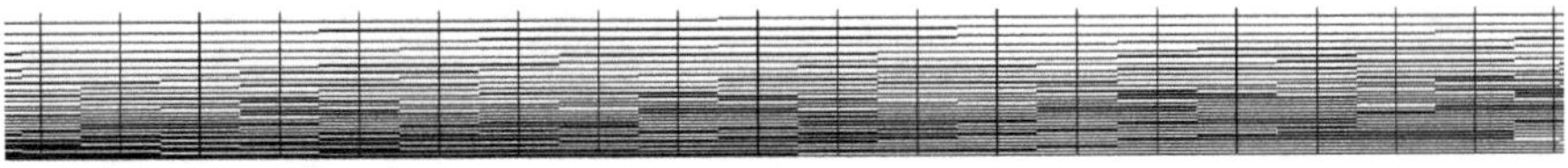

Figure 4.3. *Mesh*

– *Visualization*

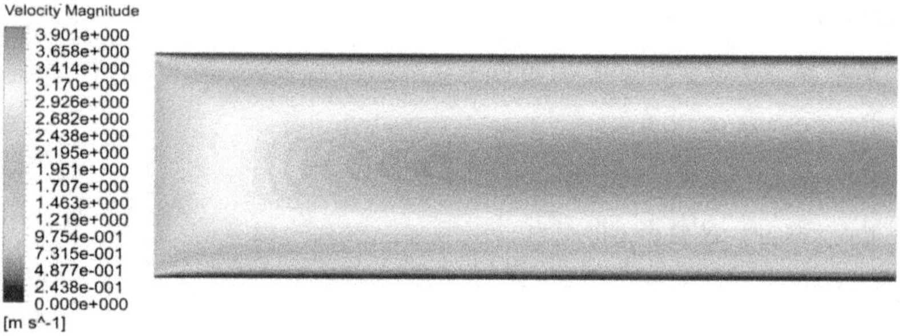

Figure 4.4. *Velocity distribution throughout the pipe. For a color version of this figure, see www.iste.co.uk/ledoux/fluid.zip*

Figure 4.5. *Static pressure contour. For a color version of this figure, see www.iste.co.uk/ledoux/fluid.zip*

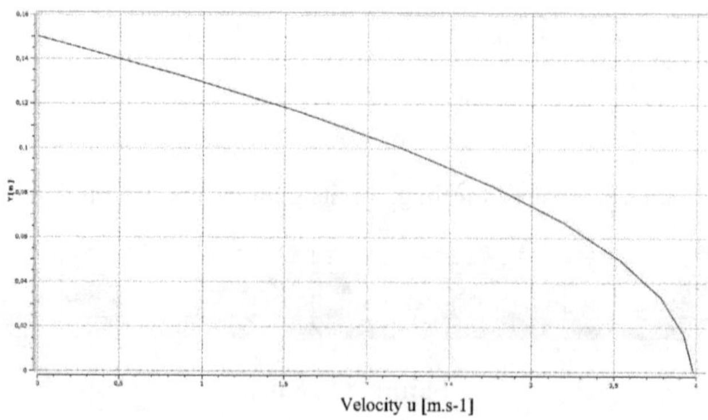

Figure 4.6. *Velocity profile at the pipe outlet in laminar regime*

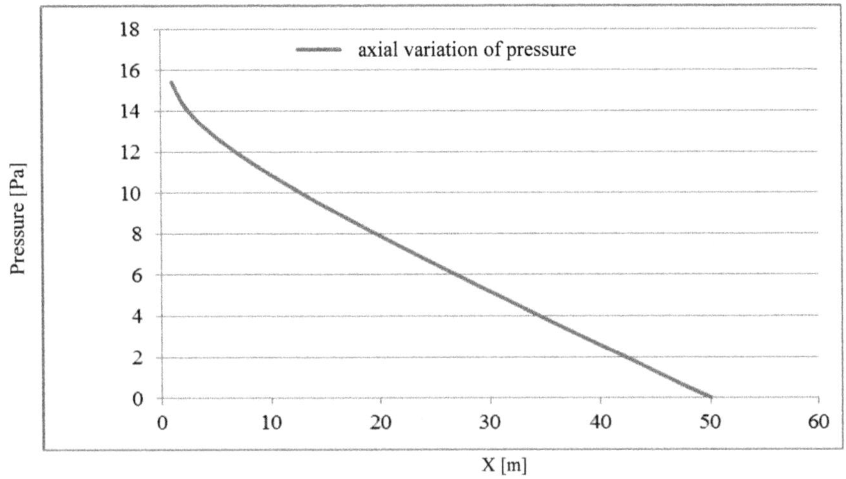

Figure 4.7. *Axial variation of pressure*

Let us now consider a turbulent flow in a pipe (Re = 10,000).

Given its performances in comparison to other turbulence models, we have chosen the k–ω SST model for modeling the turbulences. In this case, the turbulence kinetic energy can be displayed, this being needed for predicting the flow of fluid in its turbulent regime.

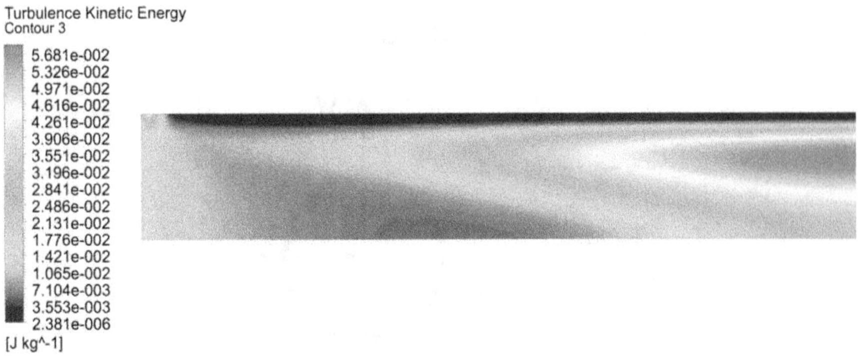

Figure 4.8. *Distribution of the turbulence kinetic energy. For a color version of this figure, see www.iste.co.uk/ledoux/fluid.zip*

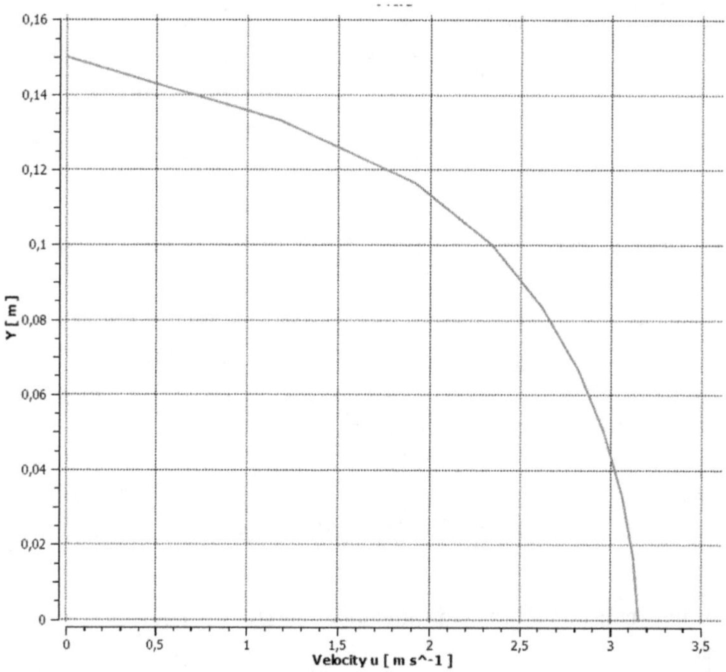

Figure 4.9. *Profile of velocity at the pipe outlet in turbulent regime*

EXAMPLE 4.2 (Laminar flow around a cylinder).–

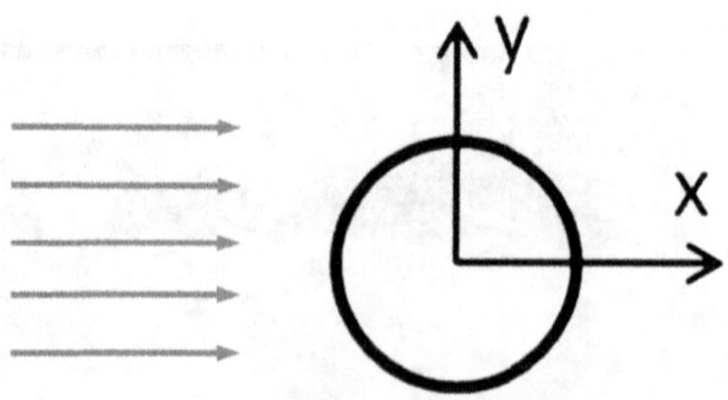

Figure 4.10. *Laminar flow around a cylinder*

This example treats the problem of the flow of an incompressible viscous fluid such as the air around a cylinder. This is one of the classical problems of fluid mechanics. The flow is characterized by the Reynolds number Re. If the flow is incompressible, this number facilitates the distinction between various regimes. To simplify the problem, we suppose the flow is laminar (Re = 80) $\mu = 0{,}05$ kg.(m s)$^{-1}$.

The velocity downstream of the cylinder is given by $V_{inlet} = 2$ m.s^{-1}.

D = 2 m (D: diameter of the cylinder)

We are interested in solving this problem under steady and transient regimes.

Solution:

a) *Geometrical model setup and definition of the physical conditions*

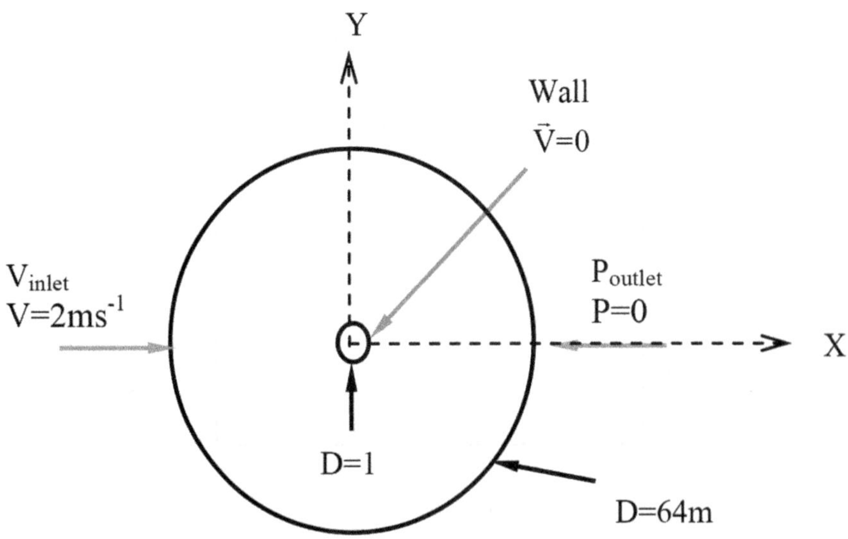

Figure 4.11. *Presentation of the computational domain*

b) *Mesh*

Let us consider an "O-grid" structured mesh [RAC 11] that is compatible with this type of problem.

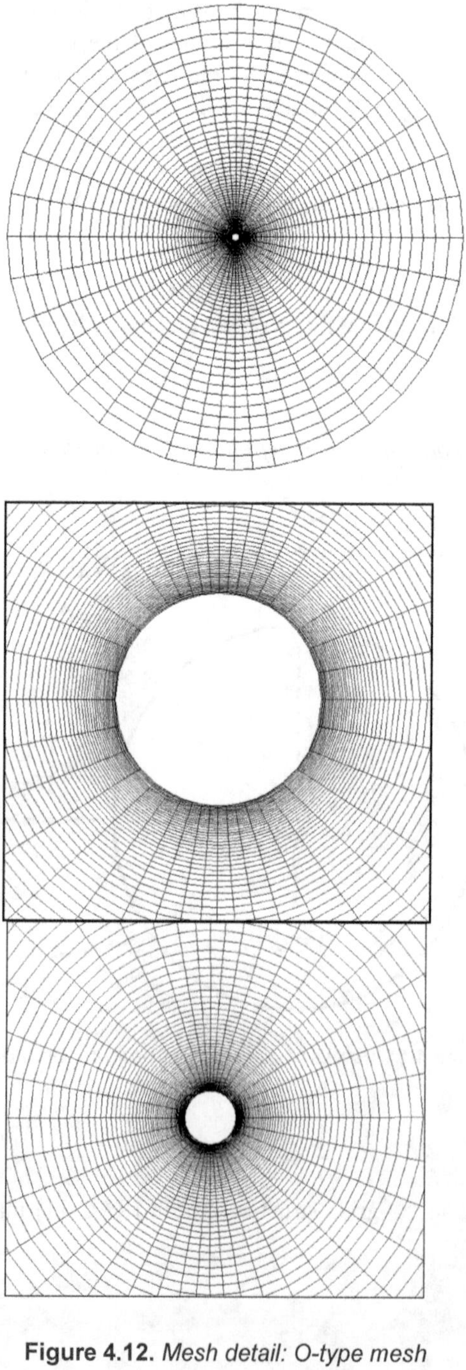

Figure 4.12. *Mesh detail: O-type mesh*

c) *Visualization*

Case 1: Steady flow

Figure 4.13. *Static pressure contour. For a color version of this figure, see www.iste.co.uk/ledoux/fluid.zip*

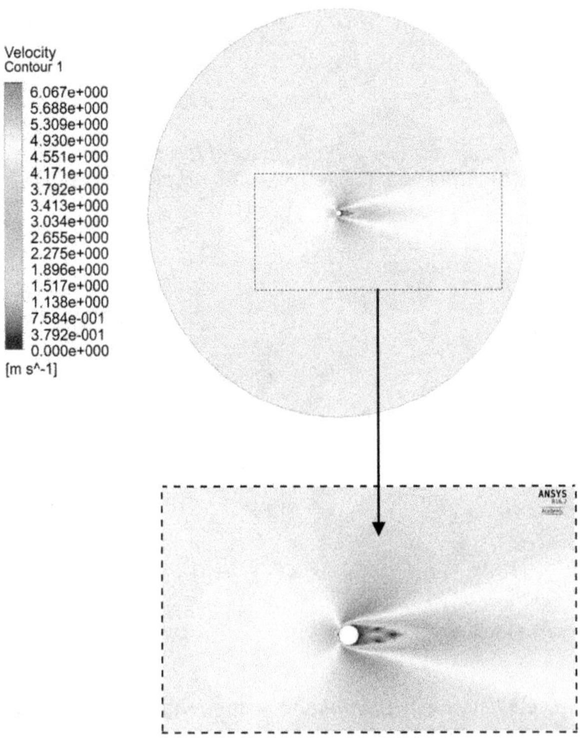

Figure 4.14. *Velocity contour around the cylinder. For a color version of this figure, see www.iste.co.uk/ledoux/fluid.zip*

The drag coefficient C_D=1,161989 is shown as:

Zone	Forces (n) Pressure	Viscous	Total	Coefficients Pressure	Viscous	Total
cylindrewall	13.509774	4.2831966	17.792971	0.88227094	0.27971895	1.1619899
Net	13.509774	4.2831966	17.792971	0.88227094	0.27971895	1.1619899

Case 2: Unsteady laminar flow

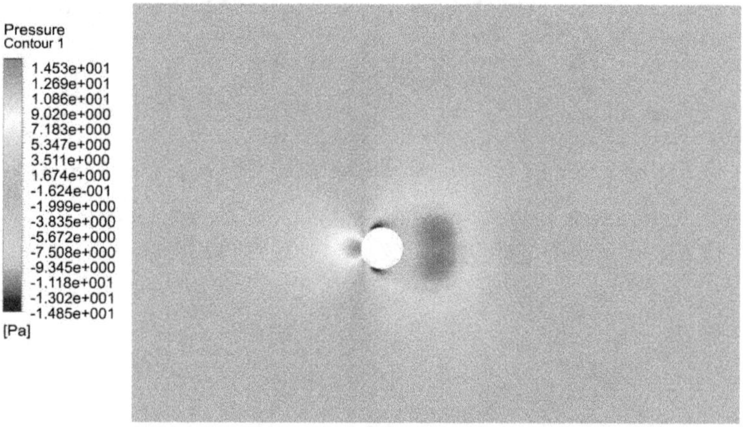

Figure 4.15. *Pressure contour around the cylinder. For a color version of this figure, see www.iste.co.uk/ledoux/fluid.zip*

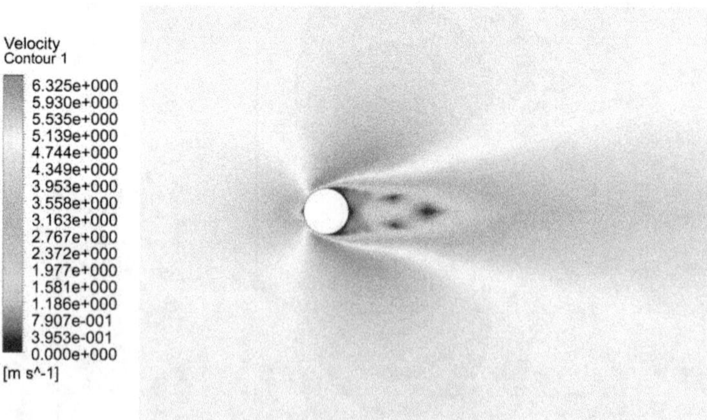

Figure 4.16. *Velocity distribution around the cylinder. For a color version of this figure, see www.iste.co.uk/ledoux/fluid.zip*

In a transient regime, the time step "Δt" plays a major role in the accuracy and convergence of the solution. The effect of time step size on the lift coefficient as a function of time is shown in Figure 4.15. It can be noted that when the time step Δt decreases, the fluctuations of the lift coefficient are more and more visualized. For a transient computation, it can be concluded that the time step directly influences the accuracy and quality of the result. Small variations of the time step provide good results.

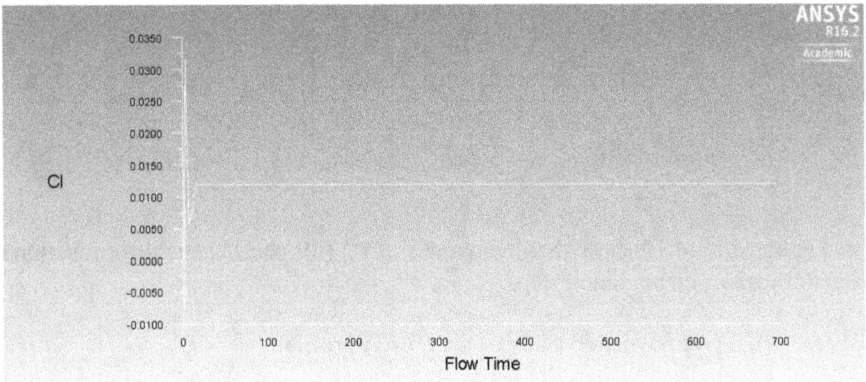

Figure 4.17. *Convergence of the solution for Δt = 1 s*

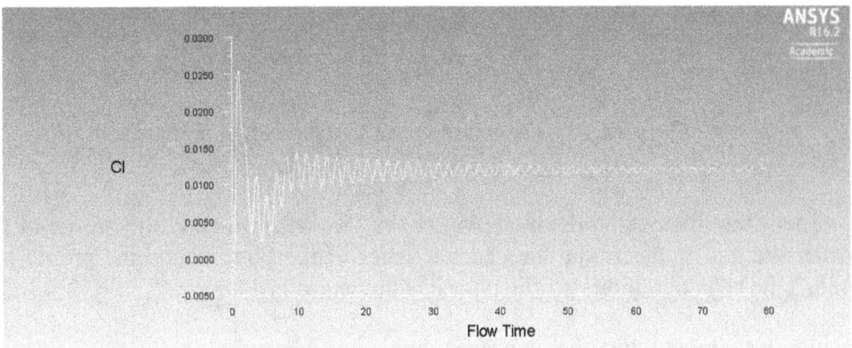

Figure 4.18. *Convergence of the solution for Δt = 0,2 s*

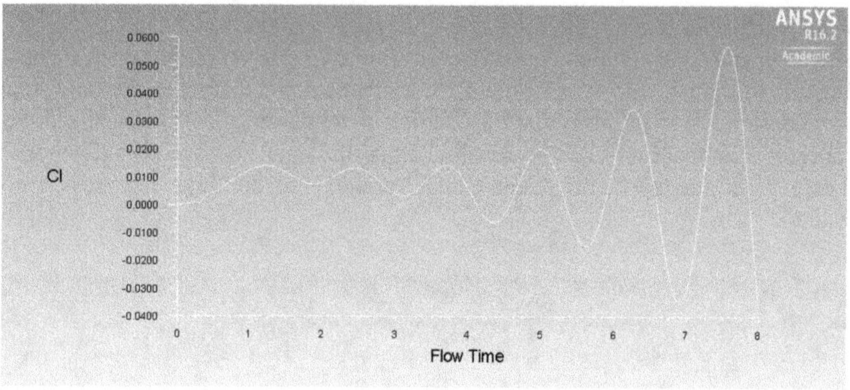

Figure 4.19. *Convergence of the solution for Δt = 0,02 s*

Figures 4.17–4.19 show the fluctuations of C_L (lift coefficients) as a function of time for various computation steps.

EXAMPLE 4.3 (Aerodynamic study of the NACA profile).–

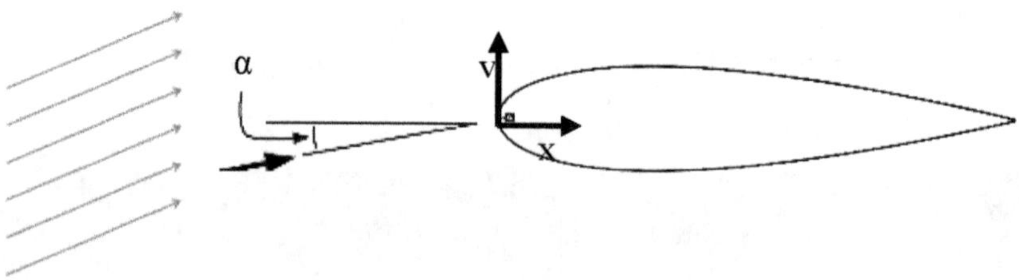

Figure 4.20. *Aerodynamic study of the NACA profile*

This study focuses on the modeling of the flow of air around the profile of the blade. We aim to determine the characteristics of the flow around the profile and predict the behavior of this profile once it is immersed in a flow.

For this purpose, the following steps should be taken:

1) Numerically compute the aerodynamic coefficients that are characteristic to the profile: lift coefficients (C_L) and drag coefficients (C_D) for various angles of attack.

These coefficients are dimensionless parameters depending on the aerodynamic profile, the Reynolds number and the angle of attack, and they can be analytically expressed as follows:

$$C_L = \frac{L}{0.5\rho A V^2} \text{ : Lift coefficient}$$

$$C_D = \frac{D}{0.5\rho A V^2} \text{ : Drag coefficient}$$

where L and D are the lift and drag forces, respectively, A is the area swept out and V is the velocity of wind.

2) Determine the distribution of pressure on the profile

3) Determine the fluid velocity contours around the profile

Solution

a) *Geometrical model setup and definition of the physical conditions*

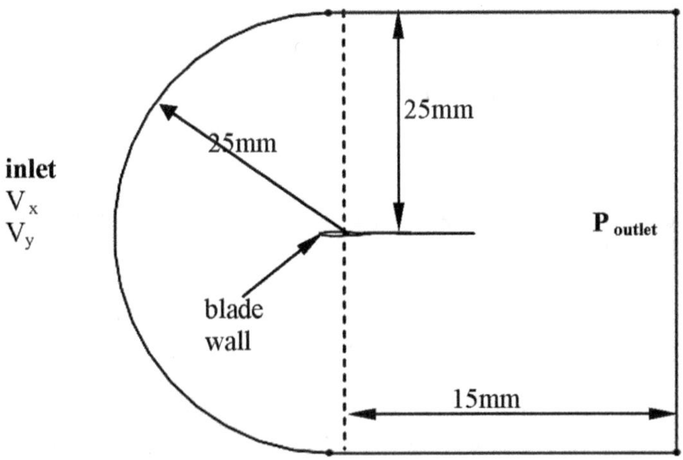

Figure 4.21. *Computational domain*

b) *Mesh*

A C-type structured mesh is adapted to this geometry. The mesh details are shown in Figure 4.22.

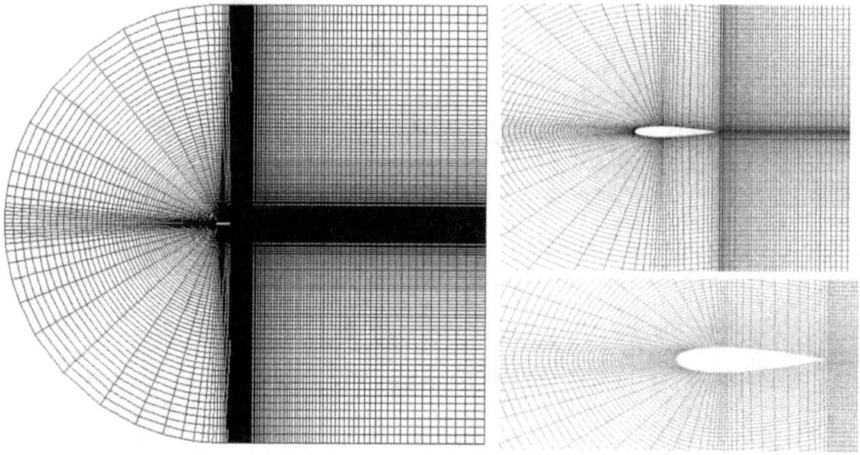

Figure 4.22. *C-type mesh*

c) *Visualization*

The objective of this simulation is to compare various values of wind velocity resulted from the variation in the angle of attack of the symmetrical profile NACA00XX.

We note that the angle of attack has a significant effect on the distribution of velocity around the blade.

The contours of velocity show that the velocities on the higher surface of the aerodynamic profile are higher than those on the lower surface. Therefore, the pressure on the higher surface is lower than that on the lower surface of the aerodynamic profile, as indicated in Figures 4.23–4.33. The pressure reaches a maximum value at the leading edge, then diminishes progressively along the aerodynamic profile. According to Bernoulli's theorem, this difference in pressure around the aerodynamic profile results in a lift force. The values of C_L and C_D, the contours of velocity and the contours of pressure are presented below for various angles of attack.

Figures 4.23–4.28 show the velocity contours for various angles of attack.

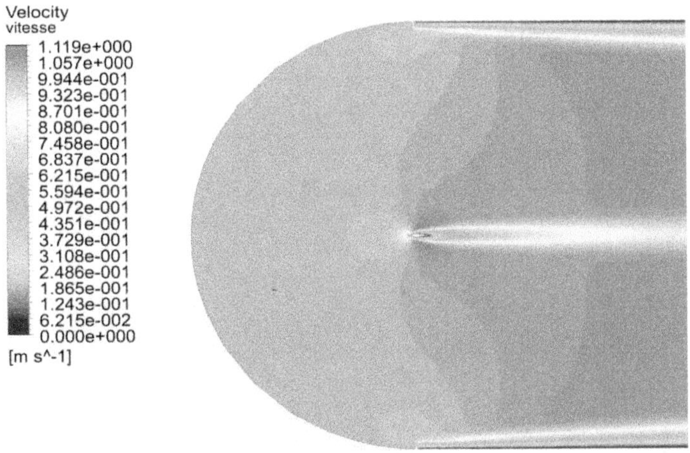

Figure 4.23. *Velocity for α = 0°. For a color version of this figure, see www.iste.co.uk/ledoux/fluid.zip*

Figure 4.24. *Velocity for α = 4°. For a color version of this figure, see www.iste.co.uk/ledoux/fluid.zip*

Figure 4.25. *Velocity for α = 8°. For a color version of this figure, see www.iste.co.uk/ledoux/fluid.zip*

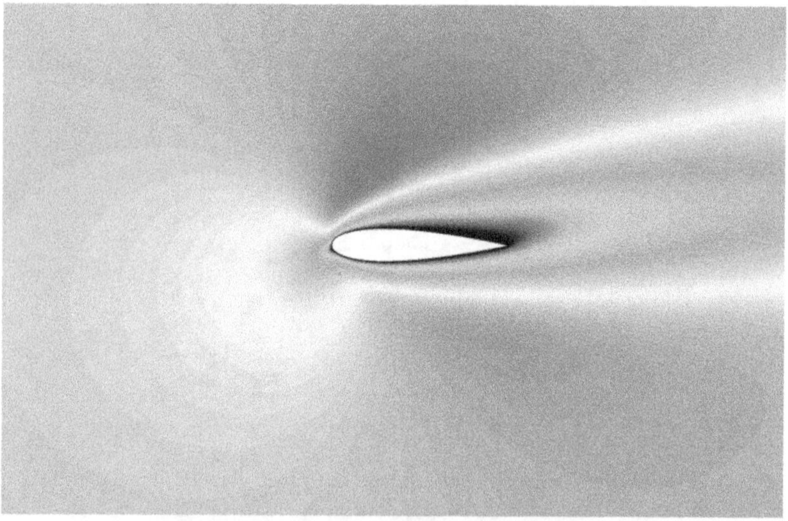

Figure 4.26. *Velocity for α = 10°. For a color version of this figure, see www.iste.co.uk/ledoux/fluid.zip*

Figure 4.27. *Velocity for α = 15°. For a color version of this figure, see www.iste.co.uk/ledoux/fluid.zip*

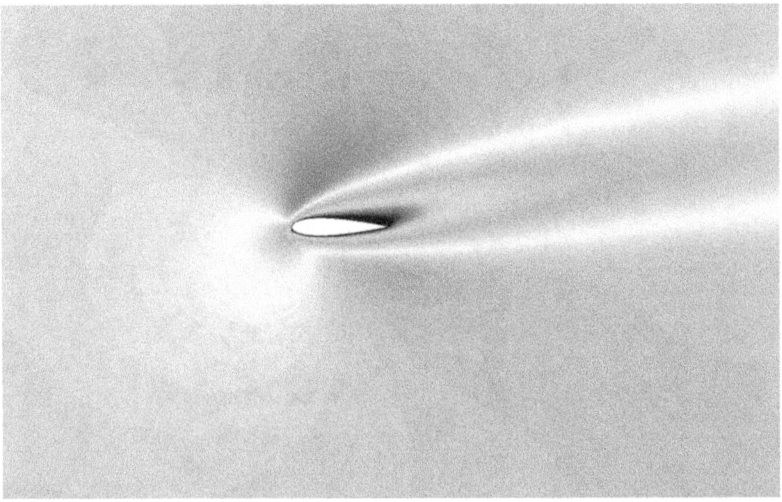

Figure 4.28. *Velocity for α = 20°. For a color version of this figure, see www.iste.co.uk/ledoux/fluid.zip*

Figures 4.29–4.33 show the pressure contours for various angles of attack.

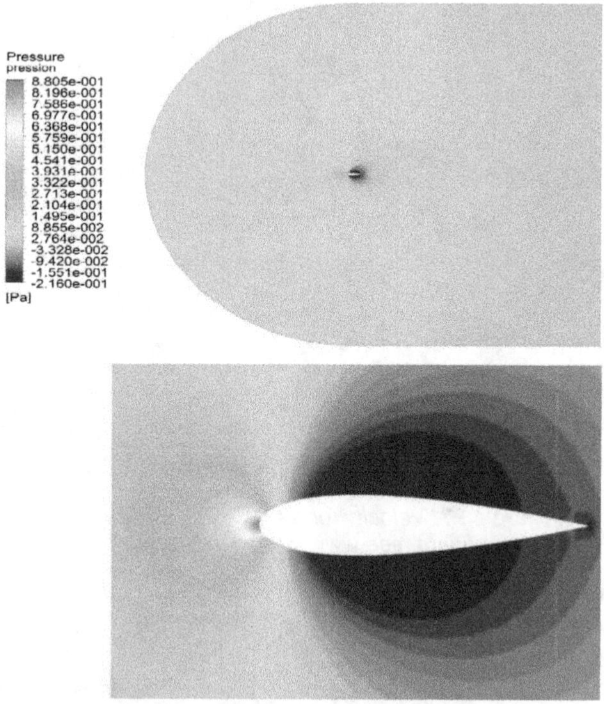

Figure 4.29. *Pressure for α = 0°. For a color version of this figure, see www.iste.co.uk/ledoux/fluid.zip*

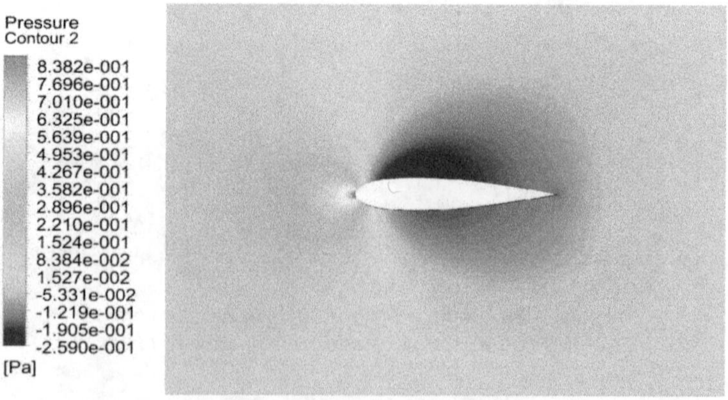

Figure 4.30. *Pressure for α = 4°. For a color version of this figure, see www.iste.co.uk/ledoux/fluid.zip*

Figure 4.31. *Pressure for α = 10°. For a color version of this figure, see www.iste.co.uk/ledoux/fluid.zip*

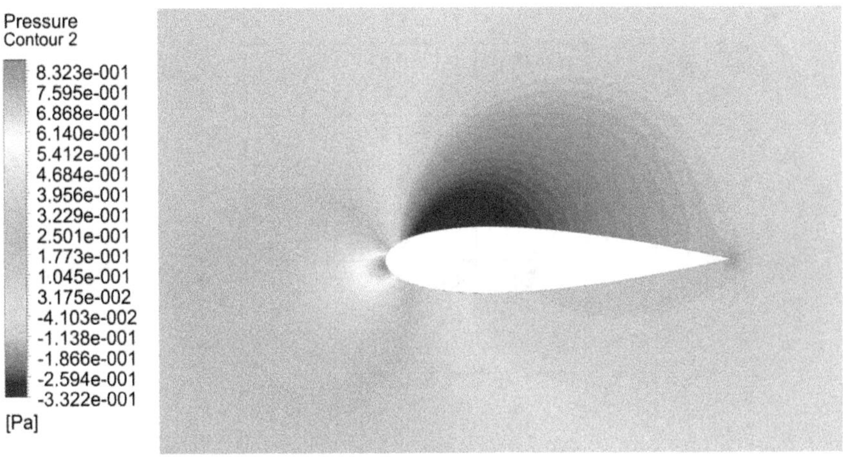

Figure 4.32. *Pressure for α = 15°. For a color version of this figure, see www.iste.co.uk/ledoux/fluid.zip*

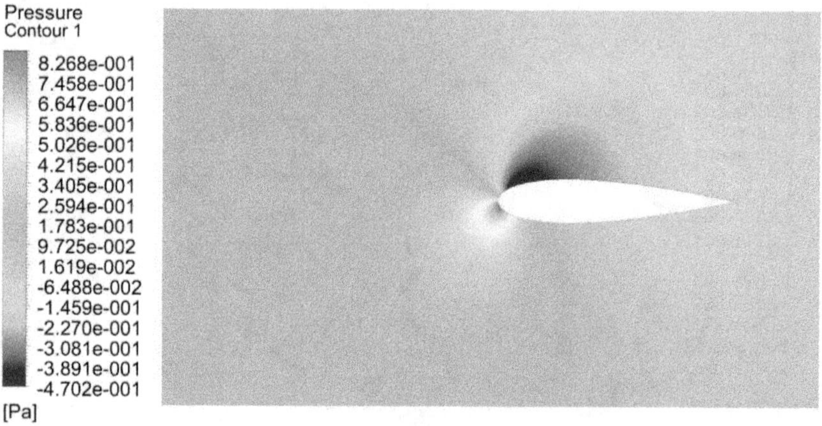

Figure 4.33. *Pressure for α = 20°. For a color version of this figure, see www.iste.co.uk/ledoux/fluid.zip*

Table 4.1 presents the values of C_L and C_D for various values of α.

$\alpha = 0°$	$C_D = 0{,}00035045$ $C_L = 4{,}14697 \times 10^{-8}$
$\alpha = 4°$	$C_D = 0{,}0005754$ $C_L = 8{,}9127 \times 10^{-5}$
$\alpha = 10°$	$C_D = 0{,}000594$ $C_L = 0{,}000217$
$\alpha = 15°$	$C_D = 0{,}0006$ $C_L = 0{,}0003$
$\alpha = 20°$	$C_D = 0{,}0006$ $C_L = 0{,}0004$

Table 4.1. *C_L and C_D*

EXAMPLE 4.4 (Plate immersed in a fluid).–

Let us consider a small rectangular plate, fixed on one side and immersed in a parallelepiped tank filled with water. The schematic representation and the data of the problem are presented in Figure 4.34 and Table 4.2, respectively.

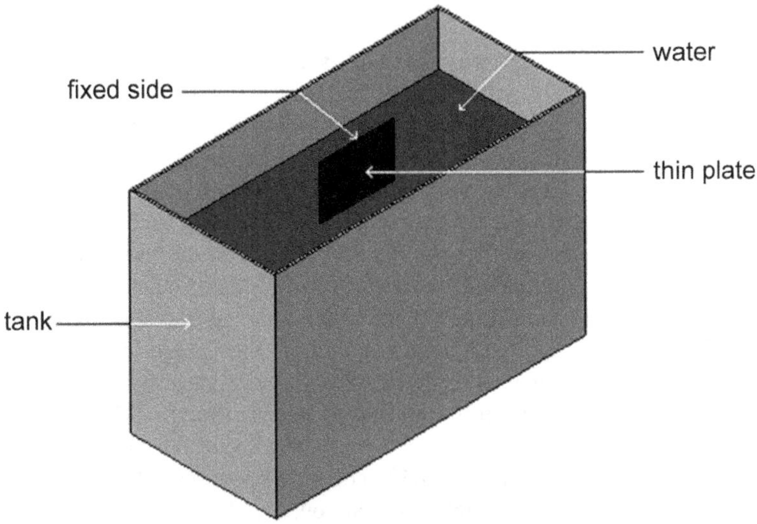

Figure 4.34. *Thin plate immersed in a tank (75% filled with water)*

	Parameter	Value
Plate	Width	72 mm
	Length	96 mm
	Thickness	1 mm
	Density	7860 kg.m^{-3}
	Young's modulus	2,1 e^{11} Pa
	Poisson's ratio	0,3
Tank	Width	14 cm
	Length	28 cm
	Height	10,4 cm
Water	Density	1000 kg.m^{-3}

Table 4.2. *Data of the problem*

a) Experimental study

The experimental study was conducted at the LOFIMS (Laboratory for Optimization and Reliability of Structural Mechanics) of I.N.S.A. (National Institute for Applied Sciences) [BEN 08] and was intended to compare the experimental result with the numerical result. The experimental setup (Figure 4.35) uses a continuous wave frequency-doubling YAG laser; the phase shift is obtained using a mirror coupled with a piezoelectric actuator, controlled by a dedicated computer card [STE 98]. The vibrations of the plate immersed in the tank with transparent walls are initiated by a second piezoelectric actuator placed outside the tank and the result can be viewed in real time on the monitor (Figure 4.36). Time-integrating interferometry [PRY 89] and high-resolution time-integrating interferometry [BOR 07] are used. The images rendered by these techniques for the first modes of the plate in the air and in full immersion are presented in Figure 3.37. The experimental study reveals that the sixth mode is affected by the variations of the immersion rate, as shown in Figure 4.38. The next stage consists in checking if the same result is obtained by numerical computation.

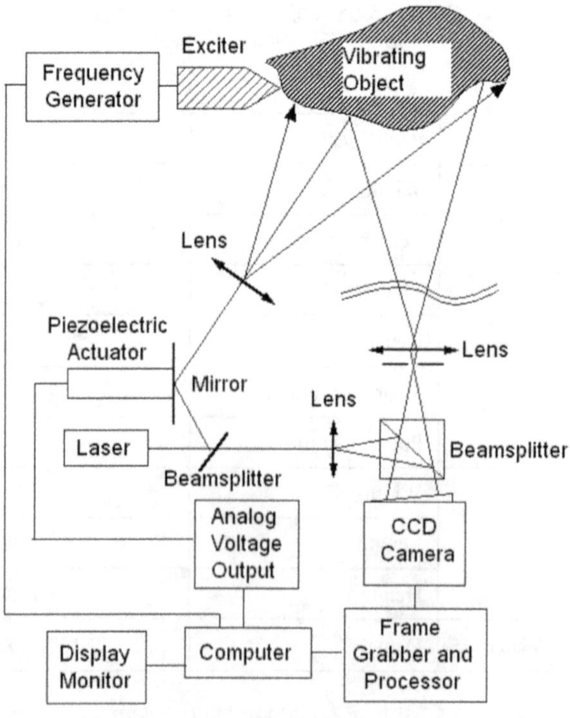

Figure 4.35. *Experimental setup*

Figure 4.36. *Image of a mode rendered by the display monitor*

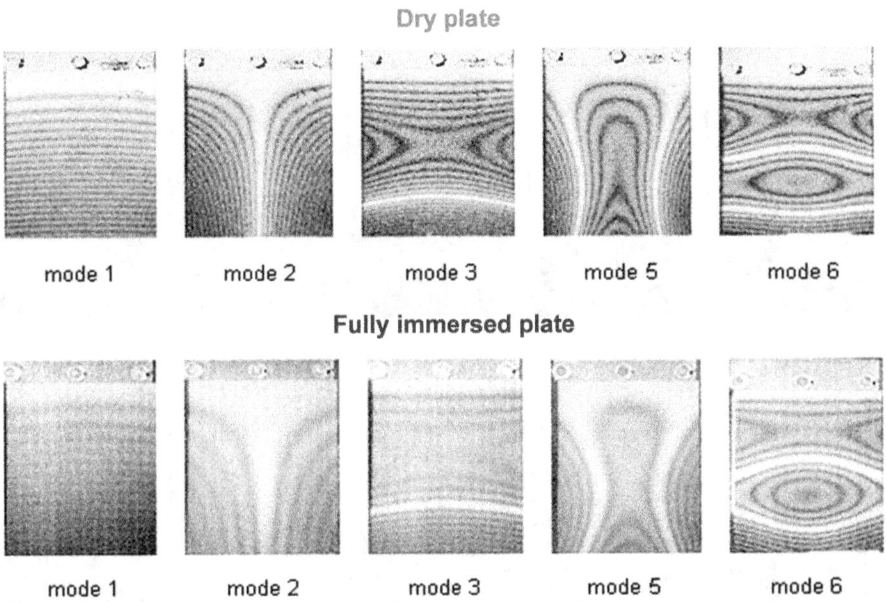

Figure 4.37. *Experimentally found deflections of the first modes*

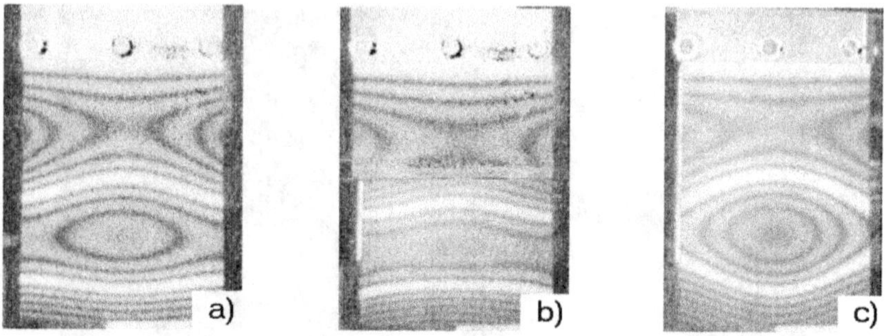

Figure 4.38. *Modification of the sixth mode as a function of the immersion rate: (a) in the air, (b) partial immersion, (c) full immersion*

b) Numerical simulation

For the numerical study, MATLAB® computation codes were used in a first stage for the dynamic computation. Figure 4.39 shows the values and nodal deflections of the first five eigen modes of the plate, calculated in the air and in full immersion in water. Figure 4.40 shows the variation of the sixth mode as a function of the rate of immersion in water.

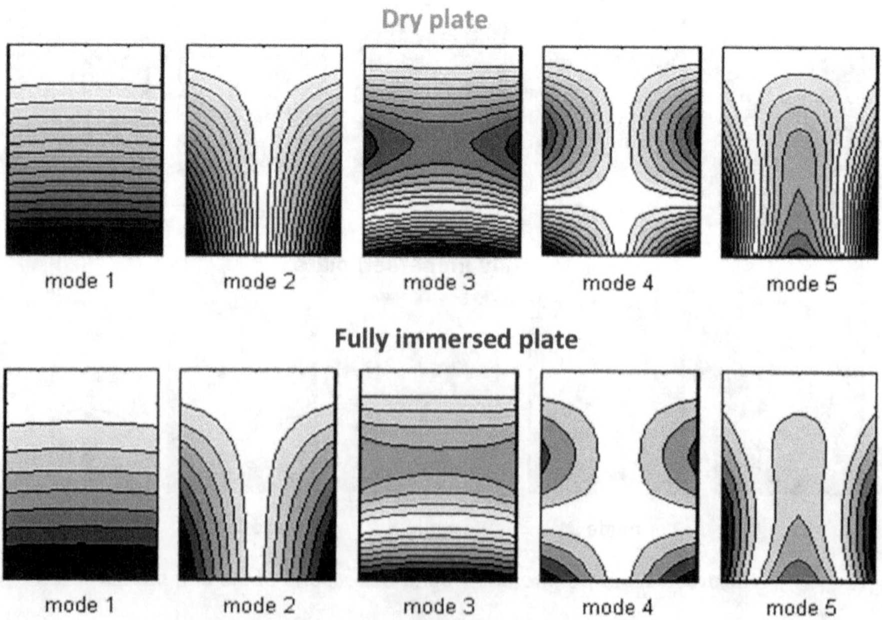

Figure 4.39. *The first five eigen modes of the plate, respectively in the air and in full immersion in water*

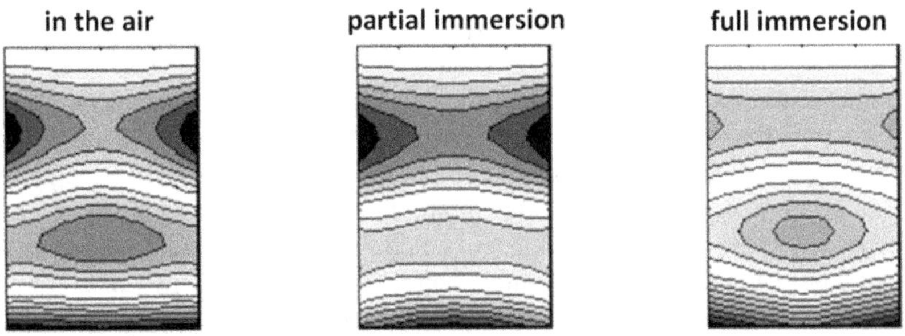

Figure 4.40. *Modification of the sixth mode depending on the immersion rate*

A comparison of the experimental results and numerical results indicates that they are quite identical. Furthermore, it can be noted that Figures 4.38 and 4.40 show the same modification of the sixth mode of the plate depending on the rate of immersion in water.

The edge of the plate is then considered excited by a frequency-depending force: $f(t)=10\sin(\omega t)$. Following this excitation, the FRF (Frequency Response Function) of the plate was calculated in the air and in full immersion for $\omega \in [0, 800\,\text{Hz}]$ (Figure 4.41).

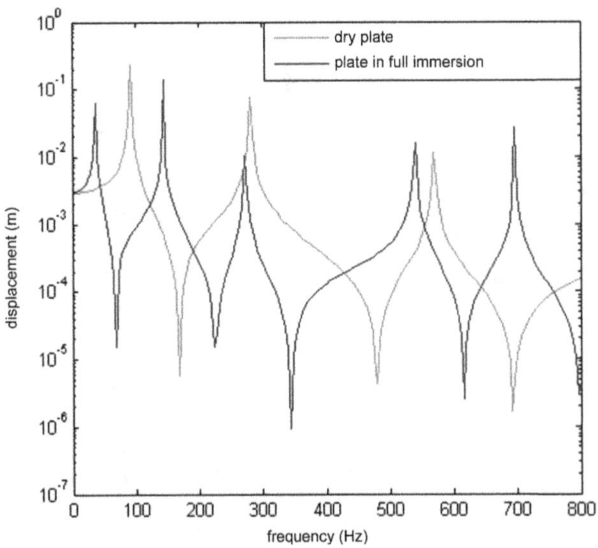

Figure 4.41. *FRF of the dry plate and fully immersed plate. For a color version of this figure, see www.iste.co.uk/ledoux/fluid.zip*

EXAMPLE 4.5 (Fluid flow and heat transfer).–

In this example, we solve a three-dimensional problem of turbulent fluid flow and heat transfer in a mixer elbow, which is a configuration used in the piping systems of power generating plants and processing industry. It is often important to predict the flow field and the temperature field in the mixing area in order to correctly define the cross-connection. A cooled fluid at 20 °C is discharged in a tube through a large inlet and is mixed with a hotter fluid at 40 °C, which flows through a smaller inlet aperture located at the level of the elbow.

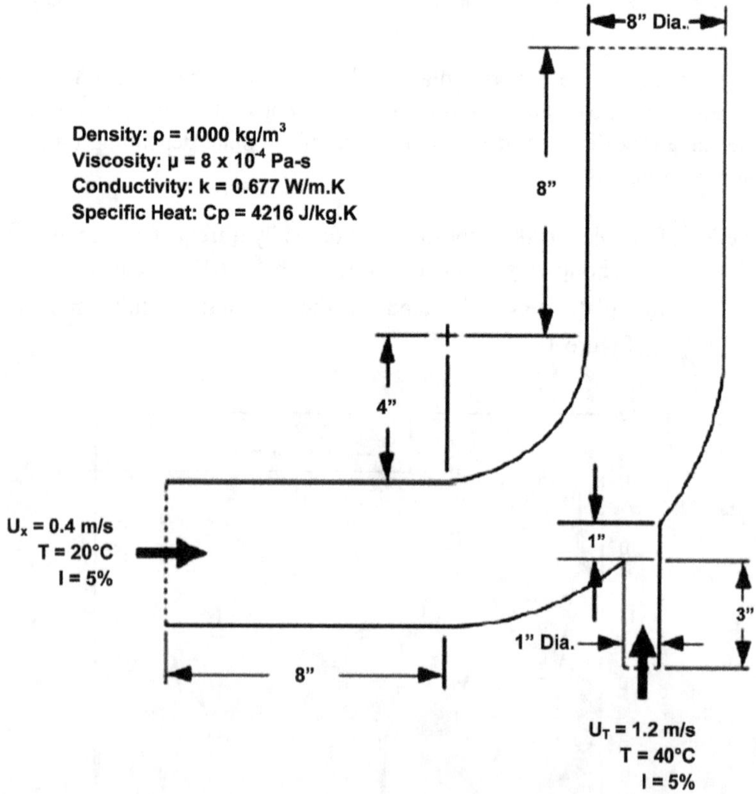

Figure 4.42. *Dimensions of the tube. For a color version of this figure, see www.iste.co.uk/ledoux/fluid.zip*

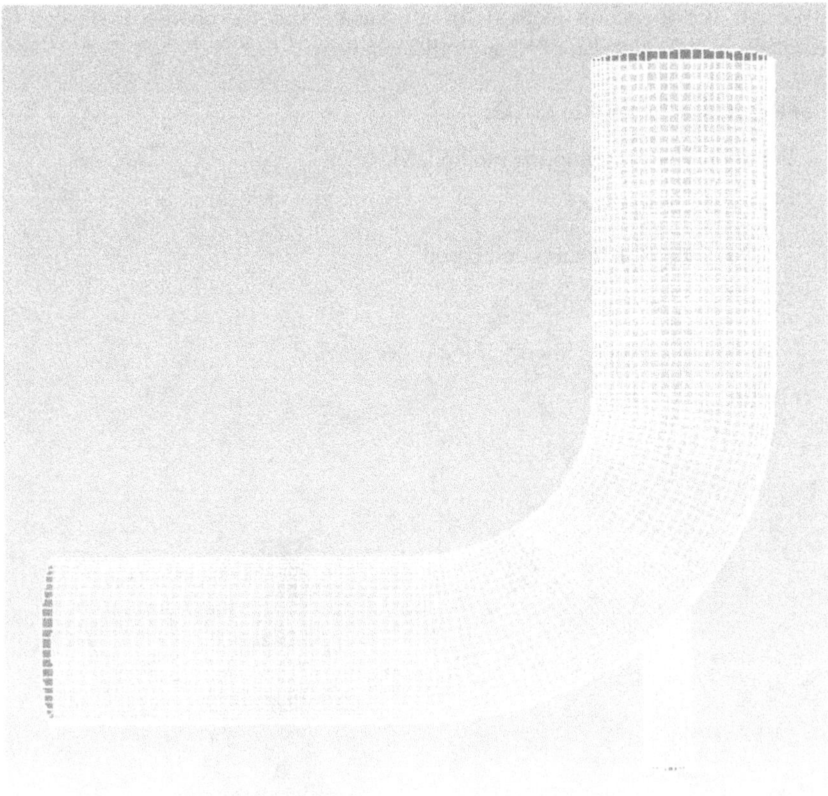

Figure 4.43. *Tetrahedral mesh (2879 nodes, 12360 elements). For a color version of this figure, see www.iste.co.uk/ledoux/fluid.zip*

1) Boundary conditions:

Outlet velocity 1: X = 0,4 m.s^{-1} Y = 0 m.s^{-1} Z = 0 m.s^{-1} Temperature = 293,15 K

Inlet velocity 2: X = 0 m.s^{-1} Y = 1,2 m.s^{-1} Z = 0 m.s^{-1} Temperature = 313,15 K

Outlet pressure: 0 Pa

We use the k-epsilon method for turbulence and we choose Enhanced Wall Treatment for the selection of Near-Wall Treatment

2) Resolution method in FLUENT:

– Pressure–velocity coupling method: SIMPLE

– Spatial discretization:

- Gradient: Least squares cell based

- Pressure: Second Order

- Turbulence Kinetic energy: First order upwind

3) Results:

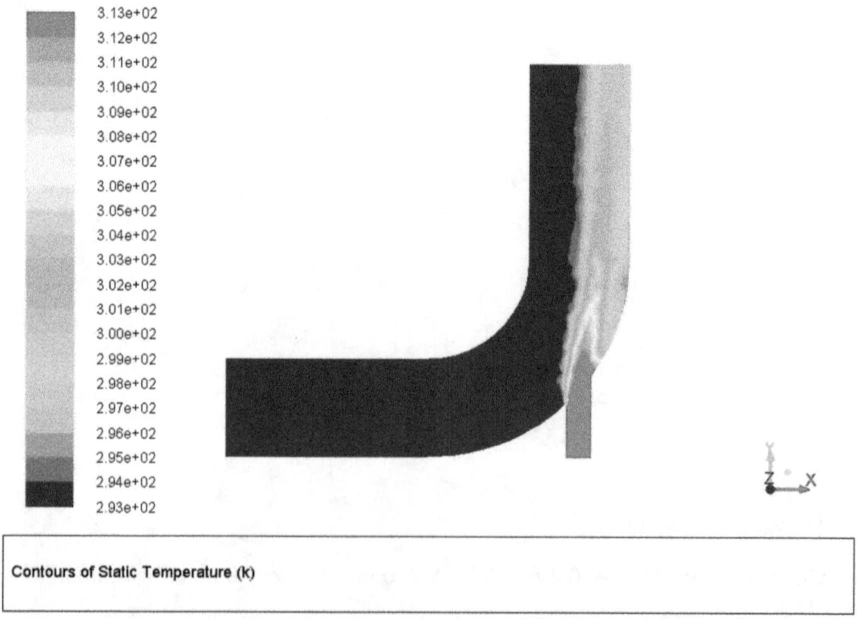

Figure 4.44. *Contours of static temperature(k). For a color version of this figure, see www.iste.co.uk/ledoux/fluid.zip*

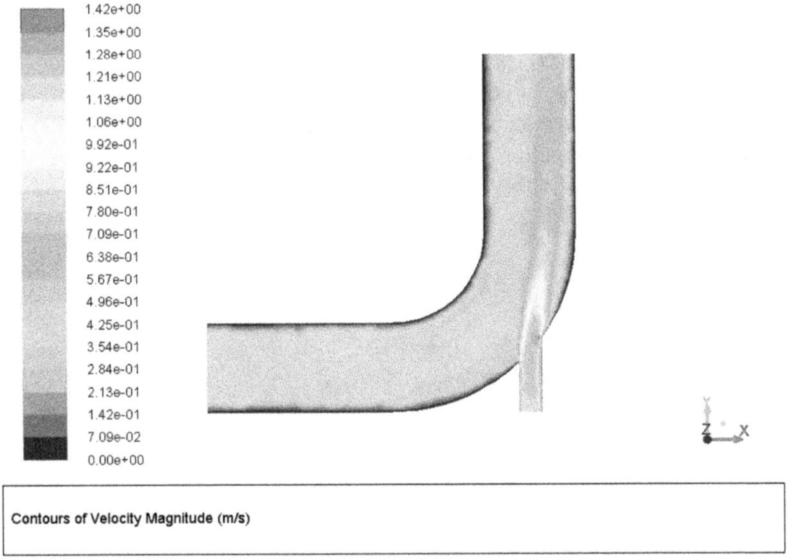

Figure 4.45. *Contour of Velocity Magnitude (m/s). For a color version of this figure, see www.iste.co.uk/ledoux/fluid.zip*

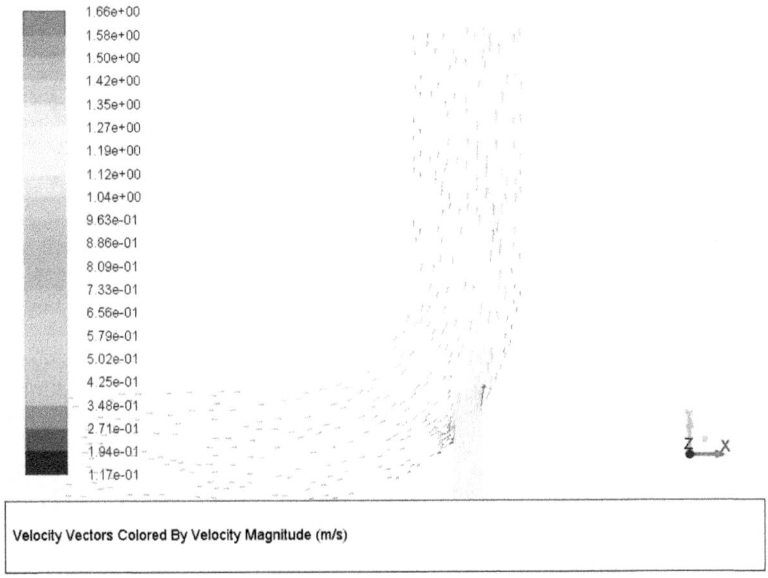

Figure 4.46. *Velocity vectors. For a color version of this figure, see www.iste.co.uk/ledoux/fluid.zip*

4.9. Conclusion

In this chapter, we have presented the CFD equations and the fluid–structure coupling equations. They have been discretized by means of the finite element method, which has led to matrix systems allowing the numerical resolution of CFD and fluid–structure interaction problems. We have also presented five different examples solved with the ANSYS FUENT software.

5

Numerical Simulation of a Vertical-axis Wind Turbine

5.1. Introduction

In this chapter, we present a performance evaluation of the straight-bladed Darrieus wind turbine. Numerical computations are compared with the experimental results found elsewhere.

A parametric study is conducted to assess the aerodynamic performances and their effect on the dynamic behavior of this wind turbine.

This chapter focuses on two-dimensional (2D) numerical computations, in which the flow variations in the vertical plane are neglected and the blade tip losses are not taken into account. The distribution of wind velocity is considered uniform at the inlet and the blades are considered straight. The performance of a straight-bladed Darrieus wind turbine is assessed by numerical computations. The proposed numerical scheme is validated by comparing the results obtained with the experimental results found elsewhere [CAI 11].

5.2. Construction of the rotor geometry and definition of the computational domain

In this chapter, we present the unsteady simulations conducted on a Darrieus-type wind turbine rotor. The Darrieus wind turbine, which is the object of this chapter, has the same dimensions as the one studied by M.R. Castelli *et al.* [CAS 11]. This choice is justified by the availability of experimental results related to this wind turbine, which is actually a study reference.

The prototype has three blades with a chord (c) of 85,8 mm, the diameter of the wind turbine being 1030 mm, which corresponds to a solidity $\left(\sigma = \frac{Nc}{R}\right)$ of 0,5, where N is the number of blades and R is the radius of the rotor. The profile used is an NACA0021 profile.

NACA 00XX Profile

The NACA (National Advisory Committee for Aerodynamics) defines a standardization of profiles based on a numbering system that highlights the main characteristic quantities. Let us consider the NACA four-digit family of profiles: the NACA four-digit series, in which the first digit indicates the maximum relative camber, the second specifies the position of this camber and the last two numbers denote the maximum relative thickness. All these numbers are given in percentage of chord.

This study focuses on the symmetrical profile NACA 00XX with a geometrical shape that can be determined with the following expression [5.1]:

$$y = \frac{t}{0.2} c \left[0.2969 \times \sqrt{\frac{x}{c}} - 0.1260 \left(\frac{x}{c}\right) - 0.3516 \left(\frac{x}{c}\right)^2 + 0.2843 \left(\frac{x}{c}\right)^3 - 0.1015 \left(\frac{x}{c}\right)^4 \right] \quad [5.1]$$

where c is the chord of the blade, x is the position along the chord, from 0 to c, y is half of the thickness for a given value of x and t is the maximum thickness expressed as fraction of the chord.

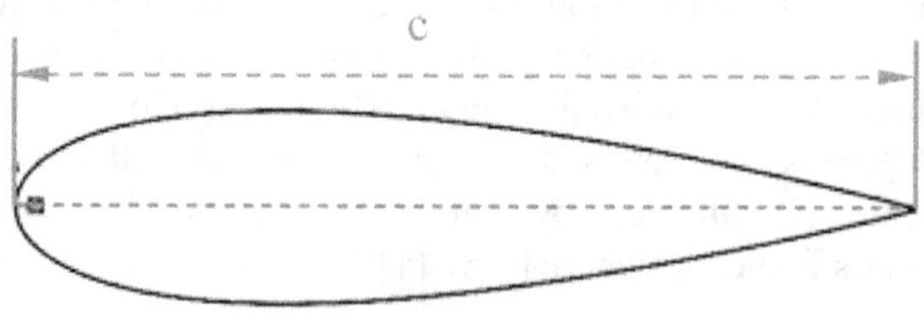

Figure 5.1. *NACA 0021 profile*

The fluid domain under consideration is rectangular, with its borders being sufficiently distant from the rotor for the solution to be free of influences exerted by the boundary conditions (length = 39D, width = 20D).

The computational domain is divided into two parts: a fixed outer domain and a slipping circular domain encompassing the rotor of the wind turbine and allowing the simulation of its rotation. The two sub-domains are separated by a slipping interface that assures flux continuity and whose mass and momentum conservation are imposed. Moreover, the rotating domain includes three mobile blade areas that have the same rotation velocity as the rotating domain. Each blade is characterized by a symmetrical profile of the type NACA 0021, and surrounded by a circle with a diameter of 400 mm. These circles have no physical meaning, but their introduction facilitates mesh refinement, since the proximity of the blades requires a sufficiently refined mesh for better accuracy.

In a CFD simulation, the specification of boundary conditions represents a critical part. It is instrumental in identifying the positions of the borders specifying these variables.

In our study, a uniform velocity inlet (V = 9 m.s^{-1}) condition is imposed at the computational domain inlet and on its lateral edges (the higher and lower borders). An atmospheric pressure outlet condition is fixed downstream of the rotor. Finally, a no-slip condition is assured on the walls of the blades. The computational domain and the boundary conditions are illustrated in Figure 5.2.

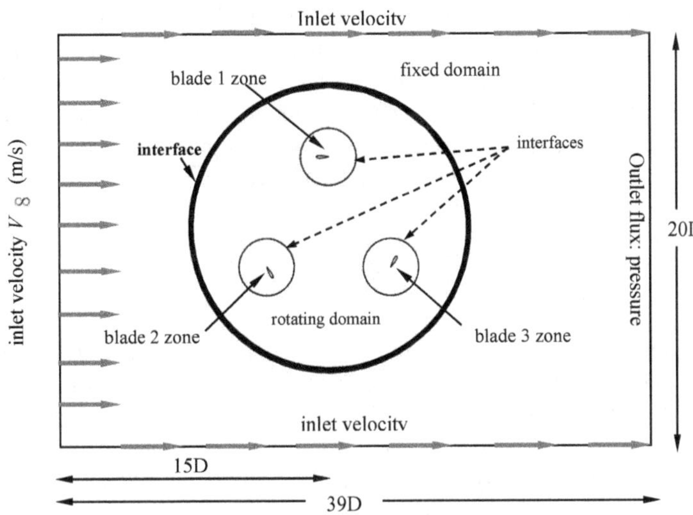

Figure 5.2. *Computational domain and boundary conditions*

5.2.1. *Mesh*

5.2.1.1. *Meshing technique*

Discretization of the computational domain is an essential phase in a CFD analysis, given how its parameters influence the accuracy of the computed solution, convergence and especially computation time. Generating a high-quality mesh is of critical importance for obtaining solid and accurate results.

There are two types of mesh: a structured mesh and an unstructured mesh. The choice of the type of grid depends on several factors, such as the numerical accuracy and the adaptability to complex geometries. Given the strong pressure and velocity gradients around the blades of the Darrieus rotor, in this model we shall use an unstructured mesh. Moreover, it is easy to generate this type of mesh on a complex geometry. In our case, mesh generation is based on the approach of the technical nomenclature QRST relying on the following rules:

– maintain high Quality of elements;

– high Resolution in the strong gradient regions;

– high Smoothing in the regions of transition from fine mesh to coarse mesh;

– minimize the Total number of elements in order to get acceptable computation times.

When implementing the finite volume method, the values of mechanical variables are computed on each control volume. Consequently, a mesh refinement is required for solving the small variations of the flux. However, the mesh refinement requirement applies only to the strong gradient regions, in order to reduce the number of elements, which in turn is a compromise between accuracy and computation time. The conclusion is that a significant mesh is characterized by high resolution in the strong gradient regions and a gradual change in the size of mesh elements from one meshing region to another while maintaining high smoothing.

5.2.1.2. *Sliding meshes*

The sliding mesh method is commonly used in the simulation of interactions between fixed and rotating parts. This method consists of interpolating the quantities at the interface between those of the sub-domains at each time step. The sliding mesh technique requires unsteady computation and generates genuinely transient computations, being generally best fitted for the simulation of flow around rotating wind turbines [FAZ 14, NOB 14, MOH 12, ALM 13].

When using the sliding mesh method, various regions of the meshed fluids rotate in discrete time steps during the unsteady computation and remain at any time adjacent to both sides of the interface.

This method leads to an optimized connectivity between two domains at the interface, while assuring the continuity of variables and the conservation of fluxes.

In our case, the blade region and the circular region move with the same rotation velocity. Four interfaces have been created between these regions: three interfaces between the rotating domain and each blade region and one interface between the fixed domain and the rotating domain.

The constructed mesh is non-uniform. It is highly refined near the interfaces and the rotor and in the proximity of the blades.

The resulting mesh is therefore an unstructured mesh (Figure 5.3) comprising around 119128 cells, out of which 85517 are distributed in the mobile regions around the Darrieus rotor.

Domain	Fixed domain	Rotating domain	Blade 1 region	Blade 2 region	Blade 3 region
Number of elements	33611	37664	16688	15316	15849

Table 5.1. *Number of elements of sub-domains*

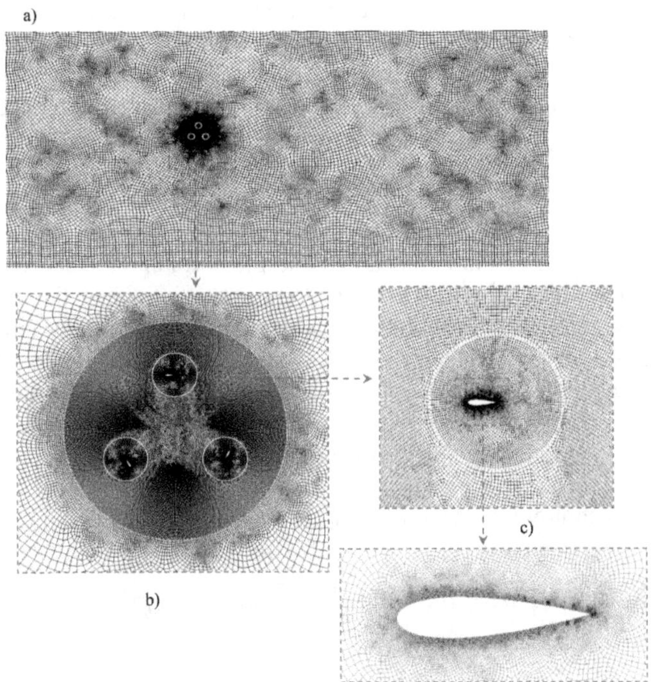

Figure 5.3. *Mesh images: general (a), around the rotor (b) and in the vicinity of the blades (c)*

5.2.2. *Discretization scheme*

Considering that the model is unsteady, discretization of the equations of conservation consists of a time discretization and a space discretization.

5.2.2.1. *Space discretization: the finite volume method*

Fluid mechanics problems are represented by nonlinear coupled partial differential equations. These equations have no analytical solutions. This is why numerical methods are used.

The CFD codes generally use four methods for the numerical resolution of partial differential equations: the finite element method, the finite difference method, the finite volume method and the spectral method. Most of the CFD programs such as the FLUENT solver rely on the finite volume method to convert the system of differential equations to an algebraic system. The main advantage of this method is that it is conservative. This is the most adequate method for the treatment of the equations used in fluid mechanics [PAY 08, CON 11]. This technique uses approximations of integrals such as the finite element method, as opposed to the finite difference method, which uses approximations of derivatives. The finite difference method can be distinguished from the finite volume method by the fact that the discretized domain consists of a set of points, while for the finite volume approach the discretized space is composed of control volumes called cells. Furthermore, this approach is better fitted for the treatment of complex problems, especially the three-dimensional (3D) ones, than the classical finite difference method.

Similarly to other resolution methods, the finite volume method comprises two important stages. The first meshing stage consists of dividing the computational domain into several cells. Each cell is associated to a point at its center where the discrete values of the variables are registered. The second discretization stage corresponds to the equations of the conservation laws being integrated over each cell. These equations are discretized by taking into account the unsteady transport equation of a scalar quantity Φ in its integral form.

For a given control volume, this equation can be written [CON 11] as:

$$\int_{v} \frac{\partial \rho \phi}{\partial t} dv + \oint \rho \phi \vec{v}.d\vec{n} - \oint \tau_{\phi} \nabla \phi \vec{v}.d\vec{n} = \oint_{v} S_{\phi}\, dv \qquad [5.2]$$

where $\vec{v}$ is the velocity vector, $\vec{n}$ is the vector normal to the surface, τ_{ϕ} is the diffusion coefficient for Φ, $\nabla \phi$ is the gradient of Φ and S_{ϕ} is the source of Φ per unit volume. The discretization of the transport equation for each control volume yields:

$$\frac{\partial \rho \phi}{\partial t} v + \sum_{f}^{Nfaces} \rho_f \vec{v}_f \phi_f \overrightarrow{A_f} - \sum_{f}^{Nfaces} \tau_\phi \nabla \phi_f \overrightarrow{A_f} = S_\phi v \qquad [5.3]$$

where:

ϕ_f	value of Φ convected through the face f;
$\rho_f \vec{v}_f . \vec{A}_f$	mass flow rate through the face f;
$\vec{A}_f$	area of face f;
Nfaces	number of faces of the cell;
$\nabla \phi_f$	gradient of Φ on face f.

When using the FLUENT solver, the values of Φ are determined at the center of the control volume, and the average face values Φ_f, corresponding to the convective terms in the integral conservation equations, are interpolated from the cell center values using an upwind discretization scheme. This scheme means that the value on the face f (Φ_f) is extrapolated from the central value Φ in the cell upstream.

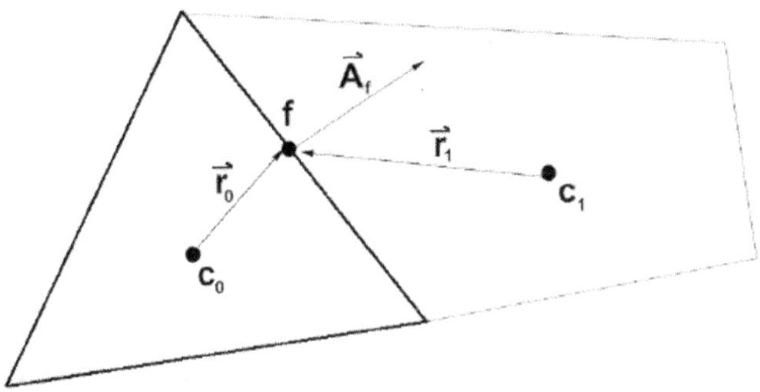

Figure 5.4. *Control volume used for the discretization of transport equations, adapted from ANSYS Inc. [CON 11]*

From the several upwind schemes available in the FLUENT solver, a second-order upwind scheme has been chosen, for accuracy reasons. For this discretization scheme, the surface value Φ_f is estimated by a second-order Taylor expansion of the central value in the cell upstream, according to the following equation [JAM 01] and [CON 11]:

$$\phi_f = \phi + \nabla\phi.\vec{r} \qquad [5.4]$$

where $\vec{r}$ is the vector of displacement of the center to the face of the upstream cell as shown in Figure 5.4. The equation requires computing the gradient $\nabla\phi$ for each control volume. This gradient is computed in FLUENT by several methods. We have chosen here the Green–Gauss cell-based method. Detailed descriptions of all these discretization schemes can be found at ANSYS Inc. [ANS 09]. The space discretization method used in this study is therefore the second-order upwind with Green–Gauss cell-based method, to further increase the accuracy.

5.2.2.2. *Time discretization*

The partial derivative with respect to time $\frac{\partial\rho\phi}{\partial t}$ has to be taken into account in the transport equation. The time variation of the flux can be expressed as a function of space variables:

$$\frac{\partial\phi}{\partial t} = F(\phi) \qquad [5.5]$$

F introduces the space discretization.

The second-order time discretization used in ANSYS FLUENT is given by:

$$\frac{3\phi^{n+1} - 4\phi^{n} + \phi^{n-1}}{2\Delta t} = F\left(\phi^{n+1}\right) \qquad [5.6]$$

where n + 1 represents the variable at t + Δt, n the variable at t and n − 1 the scalar variable at t − Δt .

$F\left(\phi^{n+1}\right)$ is evaluated at n + 1, which indicates an implicit time formulation. The equation [5.6] is iteratively solved at each time step Δt before proceeding to the next step.

For the study of wind turbine rotors, the time step Δt used for the time discretization is computed as a function of the imposed angular velocity and has to be adjusted depending on the periodicity of the study apparatus.

Each rotation velocity of the Darrieus rotor has a corresponding time step, as shown in Table 5.2.

Λ	ω [rad.s^{-1}]	Δt [s]
1,435	25,07	0,0025
1,679	29,34	0,00214
2,037	35,6	0,00176
2,326	40,64	0,00154
2,634	46	0,001365
3,085	53,91	0,00116
3,293	75,54	0,00109

Table 5.2. *Relationship between angular velocity and time step*

5.2.3. *System resolution and convergence*

The FLUENT code contains two numerical models: a density-based solver and a pressure-based solver. The former has been developed for high-velocity, compressible flows, while the latter has been created for low-velocity, incompressible flows. Our case corresponds to incompressible flows; we therefore use the pressure-based solver. According to this method, the equations of conservation and continuity of the momentum are coupled in view of the computation of the pressure field.

Therefore, the algebraic system to be solved has a set of transport equations (continuity and conservation of the momentum) that are discretized and linearized for all the cells and boundary conditions. This system is iteratively solved up to the solution convergence, more precisely until the residual difference reaches a sufficiently low value. In our computations, we have carried out iterations up to residuals of the order of 10^{-5}.

Mesh	
Structure	Unstructured mesh with quadrilateral elements
Rotation method	Sliding mesh
Mathematical model	
Conservation equation	Continuity and Navier–Stokes, unsteady
Turbulence model	k–ω SST
Discretization scheme	
Space discretization	Second-order upwind
Gradient discretization	Green–Gauss cell-based
Time discretization	Second-order implicit
Resolution method	
Pressure-based	
Boundary conditions	
Inlet	Velocity inlet
Outlet	Pressure outlet
On the walls	No-slip

Table 5.3. *Parameters used in numerical simulations*

5.3. Analysis of the results

5.3.1. *Validation of the CFD model*

The FLUENT software can possibly display results under various forms, graphic or numerical. Graphic observations facilitate the study of the distribution of flux variables and the flow characterization. The velocity contours are the most important, given that the energy is directly related to the velocity distribution. Furthermore, the pressure distribution can also be observed, since the pressure gradients determine the velocity distribution.

On the contrary, this software offers the possibility to measure dimensionless coefficients of the aerodynamic forces and the torque coefficient. We are interested in computing the aerodynamic torque to determine the power. The efficiency of this wind turbine is calculated as a function of the tip speed ratio using the following formulae:

$$Ct = \frac{T}{0.5\rho A V_{\infty}^{2} R} \qquad [5.7]$$

$$Cp = \lambda Ct = \frac{P}{0.5\rho A V_{\infty}^{3}} \qquad [5.8]$$

The main objective of this study is to evaluate the performances of the straight-bladed Darrieus wind turbine by numerical simulations and to validate our CFD study for this type of rotor, by comparing the obtained numerical results with the experimental results produced by Castelli [CAS 11]. We have used the same geometry and the same flow conditions. The geometrical characteristics of the wind turbine are presented in Table 5.4.

Profile	NACA0021
c [mm]	85,8
D [mm]	1030
H [mm]	1456,4
A [m^2]	1,236
σ [-]	0,5
N [-]	3

Table 5.4. *Experimental device for the validation of the CFD model*

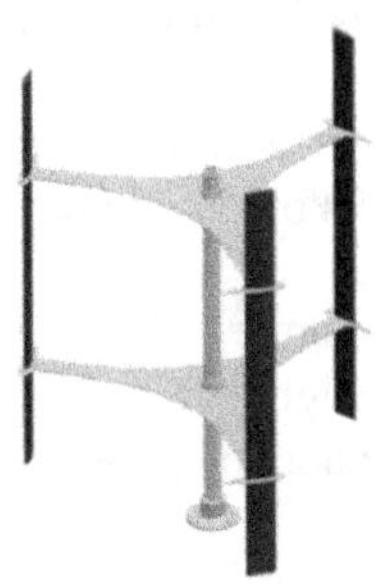

Figure 5.5. *Straight-bladed Darrieus wind turbine*

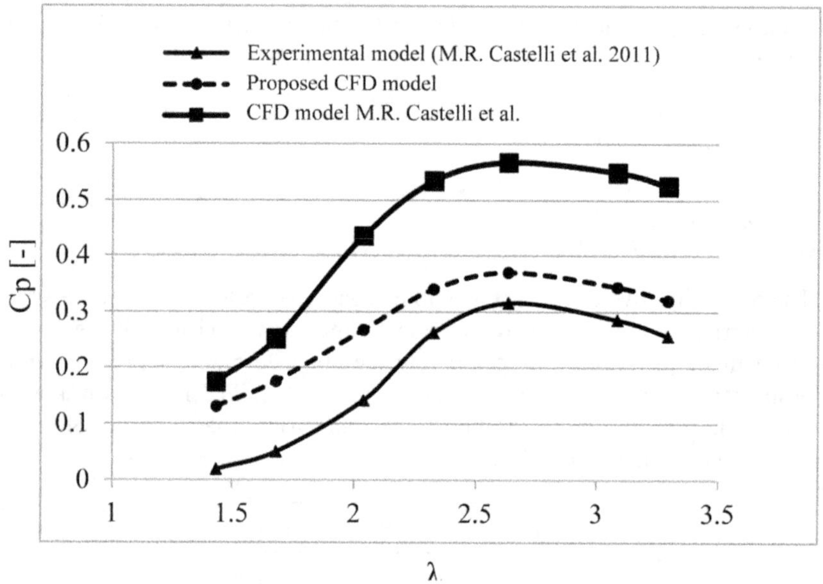

Figure 5.6. *Validation of the proposed CFD model compared with the experimental and numerical results of a Darrieus wind turbine [CAS 11]*

In order to plot the curve of the power coefficient, seven computations were launched for various rotation velocities, varying from 25,07 to 75,54 rad.s^{-1} for a wind velocity fixed at 9 m.s^{-1}. The computations have therefore been performed over an interval of tip speed ratio ranged between λ=1,435 and λ=3,293 covering the optimal operating point.

The proposed turbulence model is validated by comparing with the experimental and numerical results of a straight-bladed Darrieus wind turbine [CAS 11].

As shown in Figure 5.6, there is a very high concordance between the CFD results and the experimental results obtained with the k–ω SST model.

The difference between the numerical and experimental values of Cp can be explained by the effect of blade height and their connections. These latter constraints strengthen the drag effect and are not comprised in a 2D CFD study.

The curve reaches its maximum (Cp=0,37) for a tip speed ratio λ=2,634 and then gets inflected for higher values of λ.

It can therefore be concluded that the numerical computations are valid.

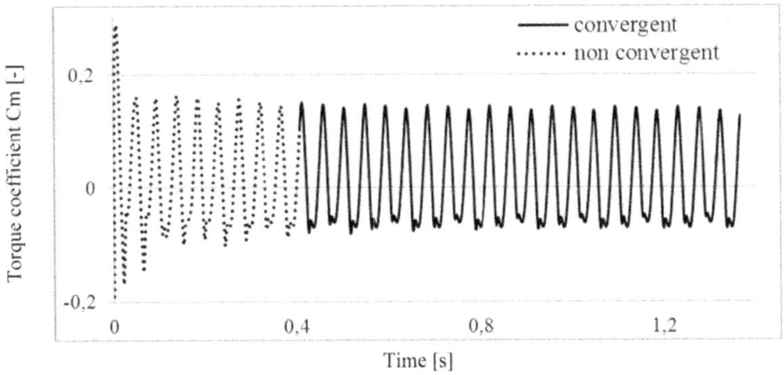

Figure 5.7. *Transient evolution of the torque coefficient during 10 rotation cycles*

Figure 5.7 presents the numerical results of the torque coefficient *Cm* as a function of time after 10 full rotation cycles of the wind turbine. This curve is plotted for an optimal tip speed ratio ($\lambda = 2{,}634$).

This plot shows an irregularity of the solution for the first three rotation cycles, then the solution stabilizes after the seven following rotation cycles during which the torque coefficient has periodic and uniform oscillations.

The fluctuations of the torque coefficient are validated by the results found elsewhere [CAR 12, ION 14, SID 15, JON 13, FAZ 14].

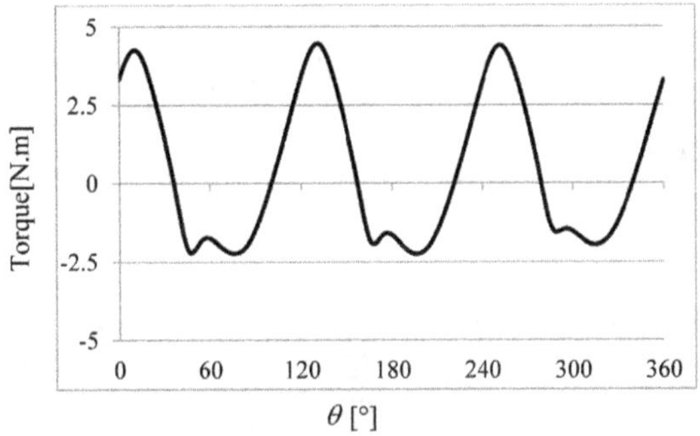

Figure 5.8. *Azimuthal evolution of the aerodynamic torque*

In terms of overall performance, the total aerodynamic torque generated by the three blades is shown in Figure 5.8. The results presented are registered during a full cycle of the rotor after stabilization of the computations.

Associations due to the rotation of the blades can be observed. It can be noted that the average overall torque is clearly positive during the whole period of revolution, which reflects good performance. In this case, the maximum overall torque is of the order of 4,48 Nm, which yields an optimal power coefficient of 0,37, corresponding to a reduced tip speed ratio of 2,634.

5.3.2. *Influence of the characteristic parameters*

The performance of the Darrieus wind turbine depends on several functional and geometrical parameters. The variation of these parameters can improve various aspects that are under research, most notably the efficiency. The focus here is on the influence of the main characteristic data of the Darrieus wind turbine, such as the tip speed ratio defined as ratio of the blade velocity at its maximum radius to the flow velocity upstream of and far away from the rotor, as well as the number of blades.

5.3.2.1. *Effect of the rotation velocity (tip speed ratio)*

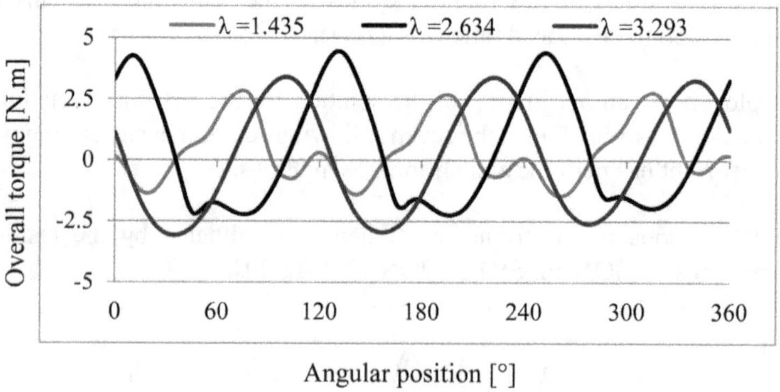

Figure 5.9. *Instantaneous torque for various tip speed ratios. For a color version of this figure, see www.iste.co.uk/ledoux/fluid.zip*

Figure 5.9 shows the aerodynamic torque generated by the three-bladed Darrieus rotor for various rotation velocities during a full cycle of the rotor (360°). The effect of rotation velocity on the aerodynamic performance of the wind turbine is illustrated in Figures 5.6 and 5.9.

It can be noted that the performance coefficient (efficiency) increases progressively with the increase in the rotation velocity until it reaches a tip speed ratio $\lambda = 2{,}634$, which corresponds to the optimal operating point. Beyond this value, the efficiency diminishes progressively, when the rotation velocity of the rotor is increased.

Moreover, it can also be noted that according to Figure 5.9, the shift from λ = 1,435 to optimal λ significantly increases the aerodynamic torque (almost doubles it), whose maximum value varies from 2,5 to 4,48 N.m, which corresponds to an increase in the power coefficient from 0,13 to 0,37. On the contrary, the shift from optimal λ to λ = 3,293 diminishes the aerodynamic torque to a less significant extent, the maximum value being reduced to 3,42 N.m. (efficiency also decreases from 0,37 to 0,32).

It can also be noted that the increase in the rotation velocity leads to an increase in the fluctuations of the aerodynamic torque.

It can be concluded that for a given geometry (constant velocity and solidity), the tip speed ratio has a direct influence on the efficiency of the Darrieus wind turbine: when the tip speed ratio is too low, there is a decrease in efficiency caused by aerodynamic stall, while a too high tip speed ratio also leads to a decrease in efficiency due to the small incidence angle and the influence of the drag force.

A proper balance of the effects results in reaching the optimal operating point, which in our case occurs at λ = 2,634.

Figure 5.10 shows the distribution of velocity contours around the Darrieus rotor for various tip speed ratios.

At the rotor passage, the inlet velocity in the region swept by the blades is reduced, and this effect is more significant at the center of the rotor, then at larger distances downstream of the wind turbine the flow starts to stabilize at incident flow values. On the contrary, an acceleration of the air masses at the tips of the blades is registered. It can be noted that for small values of the tip speed ratio (λ = 1,435), the highest velocities are of the order of 23,19 m.s^{-1}, which generates a weak aerodynamic torque. For large values of the tip speed ratio (λ = 3,293), the maximum value of the wind velocity in the proximity of the blades is 36,80 m.s^{-1}. The amplitude of the wind velocity increases significantly when the tip speed ratio increases. This is accompanied by an increase in the resultant torque, which in turn leads to a loss of efficiency. Indeed, the highest dominant gradients of velocity (v = 39,50) are registered for a tip speed ratio λ = 2,634. It can be noted that for tip speed ratio below or above this value, the wind velocity diminishes, which leads to a decrease in performance. From the velocity contours, it can be concluded that the tip speed ratio λ = 2,634 presents the optimal operating point. These results justify the fluctuation results of the aerodynamic torques (Figure 5.9).

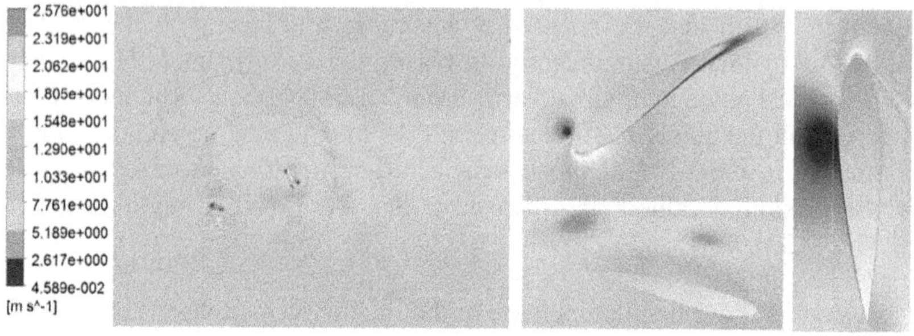

a) Low tip speed ratio λ=1.435

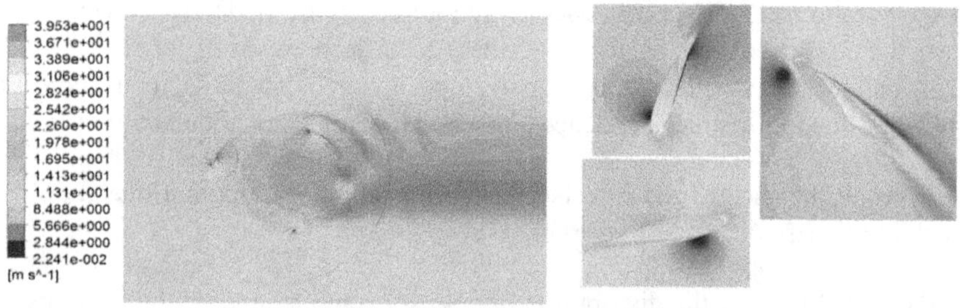

b) optimal operating point λ=2.634

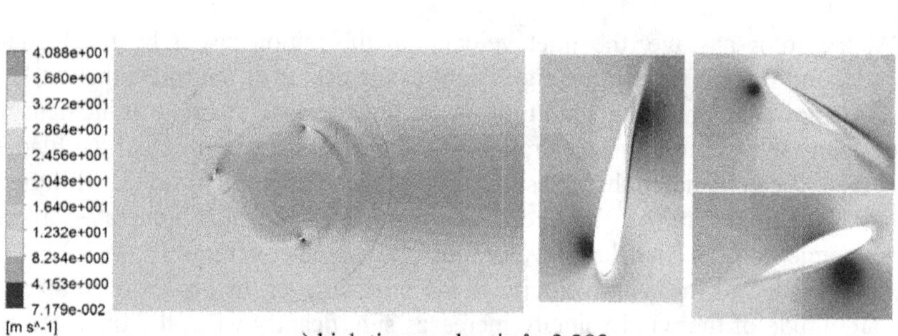

c) high tip speed ratio λ=3.293

Figure 5.10. *Velocity contours for various tip speed ratios. For a color version of this figure, see www.iste.co.uk/ledoux/fluid.zip*

5.3.2.2. *Effect of the number of blades (effect of solidity)*

Solidity is a parameter defined by Strickland [GOS 13]. On the one hand, for a given geometry (constant radius and chord), an increase in the number of blades leads to a significant increase of solidity, which reduces the maximum efficiency of the wind turbine. In this case, the increase in the number of blades could probably reduce the performance coefficient. On the other hand, reducing the solidity (with one or two blades) leads to reducing the lift force, which generates the aerodynamic torque, leading to a reduction of the efficiency of the wind turbine [GOS 13]. It is also noted that an increase in the number of blades while changing the blade chord and maintaining its radius constant does not influence the overall performance of the wind turbine.

In our case, the radius and the chord are maintained constant, while the number of blades changes, as this is the most interesting parameter.

Figure 5.11 presents the mesh details for various configurations.

The same mesh generation method was used as previously (QRLT mnemotechnic approach) with a further refinement of the mesh around the blades of the wind turbine. A rotating mesh is used for rotating the blades, and its rotation velocity is set at 46 $m.s^{-1}$.

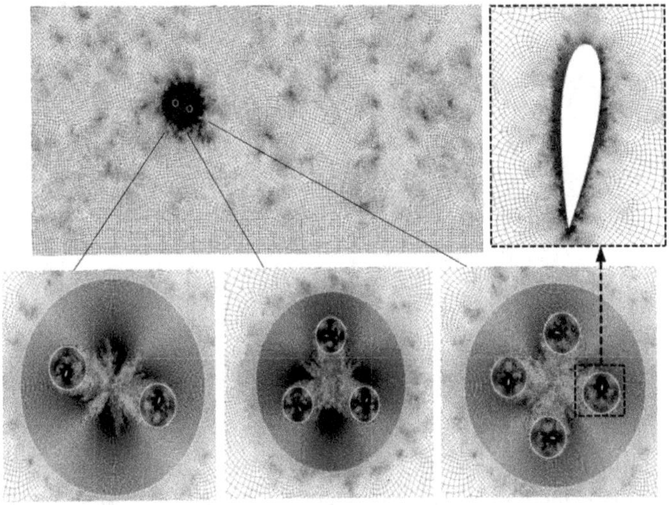

Figure 5.11. *Image of the 2D mesh model for various numbers of blades*

Figure 5.12 shows the evolution of the overall aerodynamic torque as a function of its angular position (one rotating cycle) for various numbers of blades. It is noted that the frequency of oscillation of the aerodynamic torque increases with the increase in the number of blades, since the number of periods increases during a full rotation cycle of the rotor.

Moreover, it can be noted that the aerodynamic torque generated by the various configurations has periodic fluctuations, which decrease when the number of blades increases. Nevertheless, the increase in the number of blades is apparently detrimental to the aerodynamic performance, while it facilitates the drag force caused by the increase in the number of connecting shafts. It can therefore be concluded that the number of blades has a significant effect on the efficiency of the wind turbine.

Furthermore, the decrease in the number of blades from three to two and the increase from three to four reduces the mean value of the torque coefficient (C_t, avg) and consequently the performance diminishes. The highest maximum value of the aerodynamic torque corresponds to the three straight-bladed configuration. The results obtained by CFD simulation are in agreement with the results found elsewhere [ELS 10].

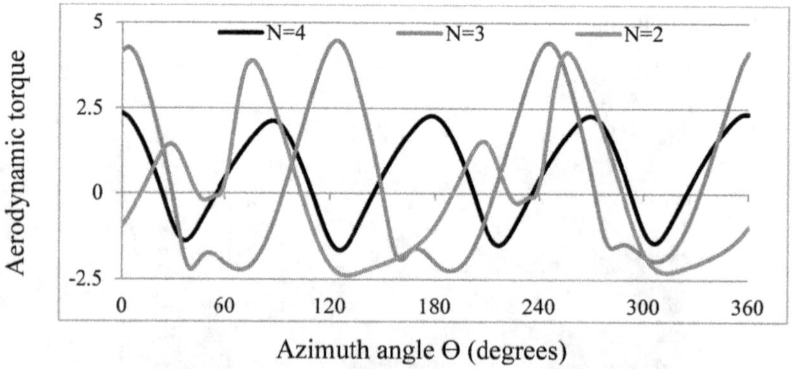

Figure 5.12. *Azimuthal evolution of the aerodynamic torque for various numbers of blades. For a color version of this figure, see www.iste.co.uk/ledoux/fluid.zip*

As previously mentioned, a three-bladed vertical axis wind turbine is more efficient than a two-bladed wind turbine in terms of aerodynamic performance, equilibrium of the structure [PAR 02], aesthetical considerations and acoustic vibration [TAN 00]. Moreover, according to Islam *et al.* [ISL 08], the three-bladed Darrieus wind turbine has the highest self-starting capacity.

In reference to the literature and to the results of the CFD simulation, we conclude that the three-bladed Darrieus wind turbine shows the highest performance aerodynamic characteristics.

	Total number of mesh elements	C_t, max	C_t, min	C_t, avg	σ
N = 2	107766	0,13	−0,075	0,0275	0,33
N = 3	119128	0,14	−0,07	0,035	0,5
N = 4	132246	0,075	−0,05	0,0125	0,66

Table 5.5. *Values of σ and C_t parameters*

The velocity and pressure fields are obtained from the resolution of Navier–Stokes equations for a 2D incompressible and turbulent flow by the finite volume method using the commercially available solver ANSYS FLUENT 15.0.

To this end, various visualizations have been carried out, particularly for the observation of velocity fields near the blades or in the rotor, given that they are strictly connected to the fluctuations of the aerodynamic torque.

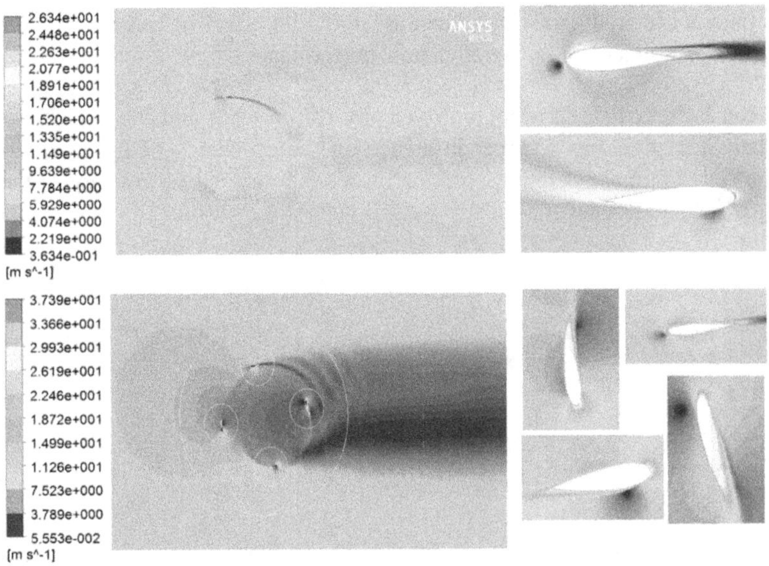

Figure 5.13. *Velocity contours for various numbers of blades. For a color version of this figure, see www.iste.co.uk/ledoux/fluid.zip*

Figure 5.13 presents the distribution of velocity around the Darrieus rotor for various numbers of blades. All these contours are registered at the optimal operating point ($\lambda = 2{,}634$).

In these various cases, a decrease in the amplitude of the wind velocity of 9 $m.s^{-1}$ can be observed upstream of the rotor as the energy is taken over by the rotor when the wind passes. Furthermore, an acceleration of the amplitude of velocity in the rotating regions around the blades can be noted with a maximum velocity of the order of 26,34 and 37,9 $m.s^{-1}$ for two and four blades, respectively (Figure 5.13), and 39,53 $m.s^{-1}$ for three blades (Figure 5.8). The highest velocity gradients further increase the production of the resultant torque. This explains why the three-bladed Darrieus wind turbine shows the highest performances. This confirms the results of the fluctuations of the aerodynamic torque for each configuration (Figure 5.12).

5.4. Conclusion

In this chapter, we evaluated the performances of a straight-bladed Darrieus wind turbine by means of numerical computations. Furthermore, the numerical scheme developed was validated, by comparing it to the experimental results obtained in the literature.

A parametric study was introduced to study the effect of certain geometric and functional parameters on the aerodynamic performances.

It can be noted that the rotation velocity of the rotor and the geometric factor (number of blades) has a direct influence on the efficiency of the Darrieus wind turbine. The optimal operating point corresponds to a rotation velocity of the order of 46 $rad.s^{-1}$ and three blades.

Appendix

General Equations of Fluid Mechanics

A.1. Review

A.1.1. *Elements of kinematics*

a) When writing the Cartesian coordinates, the following conventions apply:

The axes of Cartesian coordinates, Ox, y, z are indexed by $i = 1,2,3$

We use the Eulerian flow description.

Depending on the requirements of the presentation, the Eulerian components of velocity $\vec{V}$ are denoted by $\vec{V}(u, v, w)$ or $\vec{V}(u_1, u_2, u_3)$

b) The partial derivatives $\frac{\partial \vec{V}}{\partial t}$ should be distinguished from the full derivative $\frac{d\vec{V}}{dt}$ of velocity, which is the "real" acceleration of the fluid particle ($\frac{d\vec{V}}{dt}$ is also named hydrodynamic Lagrangian or material derivation). Expanding the expression of this acceleration, we obtain:

$$\frac{d\vec{V}}{dt} = \frac{\partial \vec{V}}{\partial t} + \left(\vec{V}.\vec{grad}\right)\vec{V} \qquad \text{[A.1]}$$

or projected on the i axis:

$$\frac{du_i}{dt} = \frac{\partial u_i}{\partial t} + u_j.\frac{\partial u_i}{\partial x_j} \qquad \text{[A.2]}$$

where $\left(\vec{V}.\overrightarrow{grad}\right)\vec{V}$ is a particular notation. The operator $\left(\vec{V}.\overrightarrow{grad}\right)$ is successively applied to each component of $\vec{V}$. In this case, $\vec{V}.\overrightarrow{grad}\,u$, $\vec{V}.\overrightarrow{grad}\,v$ and $\vec{V}.\overrightarrow{grad}\,w$ are in fact scalar products.

c) In what follows, we shall systematically use the Einstein notation. Thus:

$$T_i = \sigma_{ij}\, n_j \;\; T_i = \sigma_{ij}\, n_j \text{ is equivalent to } T_i = \sum_{j=1}^{j=3} \sigma_{ij}\, n_j \qquad [A.3]$$

The Kronecker symbol δ_{ij} is defined by:

$$\delta_{ij} = 1,\ if\ \ i = j \qquad [A.4.a]$$

$$\delta_{ij} = 0,\ if\ \ i \neq j \qquad [A.4.b]$$

A.2. Writing the laws

A.2.1. *Law no. 1: the law of continuity*

This law relates to the kinematics chapter (Chapter 3 of [LED 17])

The time variation of the mass of fluid in a domain D contained within a fixed surface S (relative to the reference system in which the equations are written) is due to the flux of matter through S. This can be written as:

$$\frac{\partial}{\partial t}\iiint_D \rho\, d\omega = -\iint_S \rho\, \vec{V}\, \vec{n}\, dS \qquad [A.5]$$

This can be converted into a local equation, the continuity equation:

$$\frac{\partial \rho}{\partial t} + div\, \rho \vec{V} = \vec{0} \qquad [A.6]$$

which can also be written as:

$$\frac{d\rho}{dt} + \rho\, div \vec{V} = \vec{0} \qquad [A.7]$$

where $\frac{d\rho}{dt}$ is the material derivative of density. This expresses the variation of ρ along the fluid. It can be noted in passing that [A.7] can be used to show that a null divergence is a necessary and sufficient condition for the fluid to be incompressible: if ρ is constant, then $div\,\vec{V} = \vec{0}$; if $div\,\vec{V} = \vec{0}$, then $\frac{d\rho}{dt}$ is null and the density is necessarily constant throughout the flow.

A.2.2. *Law no. 2: the fundamental law of dynamics*

This law is written in the form of the theorem of the center of masses for a mass of fluid contained by a domain D bounded by a surface S composed of points that follow the fluid particles on S:

$$\frac{d}{dt}\iiint_D \rho\vec{V}\,d\omega = \iint_S \vec{F}_S\,dS + \iiint_D \rho\,\vec{F}_V\,d\omega \qquad \text{[A.8]}$$

where $\vec{F}_S$ denotes the surface forces (integrating pressure) and $\vec{F}_V$ denotes the volume forces. This equation is modified, using notably a fundamental lemma, as:

$$\frac{d}{dt}\iiint_D \rho f\,d\omega = \iiint_D \rho\frac{df}{dt}d\omega \qquad \text{[A.9]}$$

This calculation leads to a set of local equations, commonly known as the equation of momentum:

$$\rho\frac{d\vec{V}}{dt} = -\,grad\;p + \vec{A}_S + \rho\,\vec{F}_V \qquad \text{[A.10]}$$

NOTE.– Throughout this text, the transitions from one equality to another are not readily obtained (vector geometry formulae are applied). The reader is recommended to consult a complete course.

The vector $\vec{A}_S$ integrates surface forces other than pressure. It is shown in this course that surface forces exerted on a small area dS of normal vector $\vec{n}\,(n_1, n_2, n_3)$

and expressed based on the resultant tension $\vec{T}$ can be expressed starting from tensions on the three surfaces normal to the three directions of the axes:

$$T_i = \sigma_{ij}\, n_j \qquad [A.11]$$

NOTE.– This can be written in a quite hybrid form as $\vec{A}_S = \overline{\overline{T}}.\vec{n}$, which expresses that $\vec{F}_S$ is the contraction of the stress tensor and the normal unit vector. The explicit use of σ_{ij} improves the legibility of expressions.

Consequently, the component i of vector $\vec{A}$ can be written as:

$$A_{Si} = \frac{\partial}{\partial x_j}\sigma_{ij} \qquad [A.12]$$

When writing the equations, we considered decomposition into two parts of the stress tensor, with components σ_{ij}, which is then expressed as:

$$\sigma_{ij} = -\,p\,\delta_{ij} + \tau_{ij} \qquad [A.13]$$

where δ_{ij} is the Kronecker symbol and p is defined as the opposite of one-third of the trace of the stress tensor $p = -\frac{\sigma_{ii}}{3}$ [A.14]

For a Newtonian fluid:

$$\tau_{ij} = \mu\left(\frac{\partial u_i}{\partial x_i} + \frac{\partial u_j}{\partial x_i}\right) + \eta\, div\vec{V}\,\delta_{ij} \qquad [A.15]$$

where μ is the dynamic viscosity (expressed in Poiseuilles or $Pa.s$ or $kg.m^{-1}.s^{-1}$ in the International System SI) and η is the volume viscosity.

$\vec{F}_V$ comprises all the volume forces.

The volume forces often derive from a potential ϕ_V:

$$\vec{F}_V = -\,grad\ \phi_V \qquad [A.16]$$

A.3. The equations: expanded form and simplifications

A.3.1. *The complete equations of fluid mechanics, also called "local equations", can be written in Cartesian coordinates*

The continuity equation:

$$\frac{\partial \rho}{\partial t}+\frac{\partial \rho u}{\partial x}+\frac{\partial \rho v}{\partial y}+\frac{\partial \rho w}{\partial z}=0 \qquad \text{[A.17]}$$

The momentum equation, projected on the three axes:

– Projection on Ox:

$$\rho\left[\frac{\partial u}{\partial t}+u\frac{\partial u}{\partial x}+v\frac{\partial u}{\partial y}+w\frac{\partial u}{\partial z}\right]=-\frac{\partial p}{\partial x}+\frac{\partial}{\partial x}\left[2\mu\left(\frac{\partial u}{\partial x}\right)\right]$$

$$+\frac{\partial}{\partial y}\left[\mu\left(\frac{\partial u}{\partial y}+\frac{\partial v}{\partial x}\right)\right]+\frac{\partial}{\partial z}\left[\mu\left(\frac{\partial u}{\partial z}+\frac{\partial w}{\partial x}\right)\right]+\frac{\partial}{\partial x}\left[\eta\left(\frac{\partial u}{\partial x}+\frac{\partial v}{\partial y}+\frac{\partial w}{\partial z}\right)\right]+\rho F_{V_x} \qquad \text{[A.18]}$$

– Projection on Oy:

$$\rho\frac{\partial v}{\partial t}+u\frac{\partial v}{\partial x}+v\frac{\partial v}{\partial y}+w\frac{\partial v}{\partial z}=-\frac{\partial p}{\partial y}+\frac{\partial}{\partial x}\left[\mu\left(\frac{\partial u}{\partial y}+\frac{\partial v}{\partial x}\right)\right]$$

$$+\frac{\partial}{\partial y}\left[2\mu\left(\frac{\partial u}{\partial y}\right)\right]+\frac{\partial}{\partial z}\left[\mu\left(\frac{\partial v}{\partial z}+\frac{\partial w}{\partial y}\right)\right]+\frac{\partial}{\partial x}\left[\eta\left(\frac{\partial u}{\partial x}+\frac{\partial v}{\partial y}+\frac{\partial w}{\partial z}\right)\right]+\rho F_{V_y} \qquad \text{[A.19]}$$

– Projection on Oz:

$$\rho\frac{\partial w}{\partial t}+u\frac{\partial w}{\partial x}+v\frac{\partial u}{\partial y}+w\frac{\partial w}{\partial z}=-\frac{\partial p}{\partial z}+\frac{\partial}{\partial x}\left[\mu\left(\frac{\partial u}{\partial z}+\frac{\partial w}{\partial x}\right)\right]$$

$$+\frac{\partial}{\partial y}\left[\mu\left(\frac{\partial w}{\partial y}+\frac{\partial v}{\partial z}\right)\right]+\frac{\partial}{\partial z}\left[2\mu\left(\frac{\partial w}{\partial z}\right)\right]+\frac{\partial}{\partial x}\left[\eta\left(\frac{\partial u}{\partial x}+\frac{\partial v}{\partial y}+\frac{\partial w}{\partial z}\right)\right]+\rho F_{V_z} \qquad \text{[A.20]}$$

A.3.2. *Application of simplifying hypotheses*

There is no general solution to these equations. Solving any problem of fluid mechanics involves the formulation of more or less restrictive hypotheses in order to simplify the equations.

NOTE.– In order to avoid resorting to such approximations, the Direct Numerical Simulation (DNS) can be used, which is very demanding in terms of computation means and time consumption.

a) The flow is often two-dimensional:

This is a case of practical application. The velocity has only two components and the momentum equation has only two projections: on Ox and Oy.

The continuity equation:

$$\frac{\partial \rho}{\partial t}+\frac{\partial \rho u}{\partial x}+\frac{\partial \rho v}{\partial y}=0 \qquad \text{[A.21]}$$

The momentum equation:

$$\rho\left[\frac{\partial u}{\partial t}+u\frac{\partial u}{\partial x}+v\frac{\partial u}{\partial y}\right]$$

$$=-\frac{\partial p}{\partial x}+\frac{\partial}{\partial x}\left[2\mu\left(\frac{\partial u}{\partial x}\right)\right]+\frac{\partial}{\partial y}\left[\mu\left(\frac{\partial u}{\partial y}+\frac{\partial v}{\partial x}\right)\right]+\frac{\partial}{\partial x}\left[\eta\left(\frac{\partial u}{\partial x}+\frac{\partial v}{\partial y}\right)\right]+\rho F_{Vx} \qquad \text{[A.22]}$$

$$\rho\frac{\partial v}{\partial t}+u\frac{\partial v}{\partial x}+v\frac{\partial v}{\partial y}$$

$$=-\frac{\partial p}{\partial y}+\frac{\partial}{\partial x}\left[\mu\left(\frac{\partial u}{\partial y}+\frac{\partial v}{\partial x}\right)\right]+\frac{\partial}{\partial y}\left[2\mu\left(\frac{\partial u}{\partial y}\right)\right]++\frac{\partial}{\partial x}\left[\eta\left(\frac{\partial u}{\partial x}+\frac{\partial v}{\partial y}\right)\right]+\rho F_{Vy} \qquad \text{[A.23]}$$

b) The flow can be incompressible:

For an incompressible fluid, the equations are simplified as follows:

The continuity equation:

$$div\vec{V}=0 \qquad \text{[A.24]}$$

The equation of dynamics:

$$\rho \frac{d\vec{V}}{dt} = - \vec{grad}\ p + \vec{T} \qquad [A.25]$$

c) When all the physical properties of the fluid are constant (including the incompressibility), and the fluid is Newtonian, we obtain the Navier–Stokes equations. When writing these equations, the kinematic viscosity resulting from the division by ρ is required.

For a two-dimensional flow, the Navier–Stokes equations are written as follows:

$$\frac{\partial u}{\partial x} + \frac{\partial v}{\partial y} = 0 \qquad [A.26]$$

$$\frac{\partial u}{\partial t} + u\frac{\partial u}{\partial x} + v\frac{\partial u}{\partial y} = -\frac{1}{\rho}\frac{\partial p}{\partial x} + 2\nu\frac{\partial^2 u}{\partial x^2} + \nu\left[\frac{\partial^2 u}{\partial y^2} + \frac{\partial v}{\partial x \partial y}\right] + \rho F_{Vx} \qquad [A.27]$$

$$\frac{\partial v}{\partial t} + u\frac{\partial v}{\partial x} + v\frac{\partial v}{\partial y} = -\frac{1}{\rho}\frac{\partial p}{\partial y} + \nu\left(\frac{\partial^2 u}{\partial x \partial y} + \frac{\partial^2 v}{\partial x^2}\right) + \frac{\partial}{\partial y}2\nu\left(\frac{\partial^2 u}{\partial y^2}\right) + \rho F_{Vy} \qquad [A.28]$$

where ν is the kinematic viscosity (expressed in myriastokes or more commonly in $m^2.s^{-1}$ in SI):

$$\nu = \frac{\mu}{\rho} \qquad [A.29]$$

It can be noted that, given the fact that the divergence of velocity is null for an incompressible fluid, the term "volume viscosity" η disappears. This is in fact a general property of all incompressible fluids, regardless of whether Newtonian or not.

d) For an incompressible, perfect fluid that is subjected to volume forces derived from a potential, we can write:

The continuity equation:

$$div\,\vec{V} = 0 \qquad [A.30]$$

The momentum equation is simplified into:

$$\frac{d\vec{V}}{dt} = -\frac{1}{\rho} \vec{grad}\, p - \vec{grad}\, \Phi_V \qquad \text{[A.31]}$$

The following identity can be used:

$$\frac{d\vec{V}}{dt} = \frac{\partial \vec{V}}{\partial t} + \vec{grad} \frac{V^2}{2} - \vec{V} \wedge \vec{rot}\, \vec{V} \qquad \text{[A.32]}$$

and finally:

$$\frac{\partial \vec{V}}{\partial t} + \vec{grad} \frac{V^2}{2} - \vec{V} \wedge \vec{rot}\, \vec{V} = -\frac{1}{\rho} \vec{grad}\, p - \vec{grad}\, \Phi_V \qquad \text{[A.33]}$$

At this point, it is required to distinguish between irrotational and rotational flows.

The curl of velocities for irrotational flows is null: $\vec{rot}\, \vec{V} = 0$

$\vec{V}$ $\vec{V}$ derives from a potential Φ (attention to $+\vec{grad}$):

$$\vec{V} = +\vec{grad}\, \Phi \qquad \text{[A.34]}$$

This potential of velocities Φ is governed by Laplace's equation:

$$div\, grad\, \Phi = \Delta \Phi = 0 \qquad \text{[A.35]}$$

where $\Delta\Phi$ is the Laplacian of Φ.

For these irrotational flows, it yields:

$$\frac{\partial \vec{V}}{\partial t} + \vec{grad} \frac{V^2}{2} + \vec{grad}\, \Phi_V = -\frac{1}{\rho} \vec{grad}\, p \qquad \text{[A.36]}$$

It can be noted that $\vec{V}$ is a gradient and the expression $-\frac{1}{\rho} \vec{grad}\, p$ has to be the gradient of a function P so that $dP = \frac{dp}{\rho}$ [A.37]

A fluid that satisfies [A.37] is said to be barotropic. In fact, it is the type of transformation undergone by the fluid in the flow that determines whether it exhibits a barotropic character or not. We mentioned a case of barotropic fluid in Chapter 2, when a perfect gas was subjected to a reversible adiabatic transformation. We have shown that in this case (equation [8.21]):

$$\frac{dp}{\rho} = \frac{\gamma}{\gamma - 1} d\left(\frac{p}{\rho}\right) \quad [A.38]$$

$$P = \frac{\gamma}{\gamma - 1} \frac{p}{\rho} \quad [A.39]$$

The incompressible flow is also a particular case of barotropic fluid for which:

$$P = \frac{p}{\rho} \quad [A.40]$$

This finally yields:

$$\vec{grad}\left(\frac{\partial \Phi}{\partial t} + \frac{V^2}{2} + P + \Phi_V\right) = 0 \quad [A.41]$$

where Φ is the potential of velocities and Φ_V is the potential of volume forces. Let us recall that this potential can associate gravity forces to other forces of various natures.

The Helmholtz function can be defined as:

$$H = \frac{\partial \Phi}{\partial t} + \frac{V^2}{2} + P + \Phi_V \quad [A.42]$$

$$\vec{grad}\, H = 0 \quad [A.43]$$

hence, the Helmholtz function is constant throughout the flow.

In a rotational flow, the curl of velocity is not consistently null. This is illustrated by the examples provided.

Let us rewrite [A.33] by evidencing the term in $\vec{V} \wedge \vec{rot}\,\vec{V}$:

$$\frac{\partial \vec{V}}{\partial t} + \vec{grad}\frac{V^2}{2} + \frac{1}{\rho}\vec{grad}\,p + \vec{grad}\,\Phi_V = \vec{V} \wedge \vec{rot}\,\vec{V} \qquad \text{[A.44]}$$

Let us consider the case of a steady flow. To summarize, we have:

– a steady flow;

– a perfect fluid;

– an incompressible fluid;

– volume forces deriving from a potential;

– then, the equation can be written as:

$$\vec{grad}\frac{V^2}{2} + \vec{grad}\frac{p}{\rho} + \vec{grad}\,\Phi_V = \vec{V} \wedge \vec{rot}\,\vec{V} \qquad \text{[A.45]}$$

Let us note that the gradient of the term on the left-hand side of the equation is normal to the surfaces defined by the equation:

$$\frac{V^2}{2} + \frac{p}{\rho} + \Phi_V = Cte \qquad \text{[A.46]}$$

[A.45] shows that the surfaces that are normal to $\vec{grad}\left(\frac{V^2}{2} + \frac{p}{\rho} + \Phi_V\right)$ are also normal to: the velocity $\vec{V}$; therefore, they contain the streamlines, which are tangent to velocities the curl of $\vec{V}$; they contain the eddy lines that are tangent to the curls.

In particular, this leads to deducing that the quantity $\frac{V^2}{2} + \frac{p}{\rho} + \Phi_V$ is constant on each streamline.

The gradient of this quantity is not null in a rotational flow (equation [A.45]). Therefore, this quantity can vary from one streamline to another, if they belong to two distinct surfaces $\frac{V^2}{2} + \frac{p}{\rho} + \Phi_V = Cst.$

Given the hypotheses accepted in this paragraph, if the above quantity is multiplied by ρ, then the quantity $\frac{\rho V^2}{2} + p + \rho\Phi_V$ is constant along the same streamline.

If the volume forces are reduced to gravity ($\Phi_V = gz$ is constant on a streamline), this proposition becomes:

In the steady flow of an incompressible, perfect fluid that is exclusively submitted to gravity forces, the quantity $\frac{\rho V^2}{2} + p + \rho gz$ *is constant along the same streamline.*

This is in fact the proof of Bernoulli's theorem.

NOTE.– There are many proofs of Bernoulli's theorem, varying from one book to another. This proof is certainly the most elegant.

A.4. Table of adiabatic flows

When treating the previous examples, we have resorted to a considerable number of formulae. As stressed before, these formulae result from a limited number of physical laws and constraints.

In the treatment of the posed problems, we are free to choose at each stage the formula corresponding to the step of the adopted reasoning. This requires some form of programming.

One can also definitively tabulate all the results yielded by formulae of practical interest. The way these can be used is abundantly illustrated by the previous examples. As we support this approach, we are recommending it here.

The tables presented here have been very easily built with a classical spreadsheet. These tables depend on the value of the polytropic exponent $\gamma = \frac{c_P}{c_V}$.

The tables below have been set up for $\gamma = 1.4$. This "standard" value is valid for the air and for several other gases to a satisfactory degree of approximation. It is worth reminding that the hypotheses adopted in this chapter serve the purpose to simplify the work required to solve the problems.

To make things easier for the reader, we have reported all the problems to the use of this value of γ. Consequently, a reader whose sole objective is to understand the question will be spared the programming task.

The programming stage cannot be avoided in case of repetitive computation, but it can readily integrate $\gamma = \frac{c_P}{c_V}$ as parameter. Having built the table for one value of the parameter, therefore for a fluid, the same programmed table can be reused for another fluid.

Below are the formulae to be programmed for each column of this table:

Compressible adiabatic flows	
p/p_0	Equation [2.43]
ρ/ρ_0	Equation [2.45]
T/T_0	Equation [2.46]
V/a_C	Equations [2.37] and [2.38]
S/S_C	Equation [2.48]
Y	Equation [2.43]

Compressible adiabatic flows y = 1,40:

M	p/p_0	ρ/ρ_0	T/T_0	V/a_c	S/S_c	Y
0,00	1,00000	1,00000	1,00000	0,00000		0,00000
0,01	0,99993	0,99995	0,99998	0,01095	57,87384	0,00447
0,02	0,99972	0,99980	0,99992	0,02191	28,94213	0,00894
0,03	0,99937	0,99955	0,99982	0,03286	19,30054	0,01341
0,04	0,99888	0,99920	0,99968	0,04381	14,48149	0,01787
0,05	0,99825	0,99875	0,99950	0,05476	11,59144	0,02233

0,06	0,99748	0,99820	0,99928	0,06570	9,66591	0,02677
0,07	0,99658	0,99755	0,99902	0,07664	8,29153	0,03121
0,08	0,99553	0,99681	0,99872	0,08758	7,26161	0,03564
0,09	0,99435	0,99596	0,99838	0,09851	6,46134	0,04005
0,10	0,99303	0,99502	0,99800	0,10944	5,82183	0,04445
0,11	0,99158	0,99398	0,99759	0,12035	5,29923	0,04884
0,12	0,98998	0,99284	0,99713	0,13126	4,86432	0,05320
0,13	0,98826	0,99160	0,99663	0,14217	4,49686	0,05755
0,14	0,98640	0,99027	0,99610	0,15306	4,18240	0,06188
0,15	0,98441	0,98884	0,99552	0,16395	3,91034	0,06618
0,16	0,98228	0,98731	0,99491	0,17482	3,67274	0,07047
0,17	0,98003	0,98569	0,99425	0,18569	3,46351	0,07472
0,18	0,97765	0,98398	0,99356	0,19654	3,27793	0,07895
0,19	0,97514	0,98218	0,99283	0,20739	3,11226	0,08316
0,20	0,97250	0,98028	0,99206	0,21822	2,96352	0,08733
0,21	0,96973	0,97829	0,99126	0,22904	2,82929	0,09147
0,22	0,96685	0,97620	0,99041	0,23984	2,70760	0,09558
0,23	0,96383	0,97403	0,98953	0,25063	2,59681	0,09966
0,24	0,96070	0,97177	0,98861	0,26141	2,49556	0,10371
0,25	0,95745	0,96942	0,98765	0,27217	2,40271	0,10771
0,26	0,95408	0,96698	0,98666	0,28291	2,31729	0,11168
0,27	0,95060	0,96446	0,98563	0,29364	2,23847	0,11562
0,28	0,94700	0,96185	0,98456	0,30435	2,16555	0,11951
0,29	0,94329	0,95916	0,98346	0,31504	2,09793	0,12336
0,30	0,93947	0,95638	0,98232	0,32572	2,03507	0,12717
0,31	0,93554	0,95352	0,98114	0,33637	1,97651	0,13094
0,32	0,93150	0,95058	0,97993	0,34701	1,92185	0,13466

0,33	0,92736	0,94756	0,97868	0,35762	1,87074	0,13834
0,34	0,92312	0,94446	0,97740	0,36822	1,82288	0,14198
0,35	0,91877	0,94128	0,97609	0,37879	1,77797	0,14556
0,36	0,91433	0,93803	0,97473	0,38935	1,73578	0,14910
0,37	0,90979	0,93470	0,97335	0,39988	1,69609	0,15259
0,38	0,90516	0,93130	0,97193	0,41039	1,65870	0,15603
0,39	0,90043	0,92782	0,97048	0,42087	1,62343	0,15942
0,40	0,89561	0,92427	0,96899	0,43133	1,59014	0,16276
0,41	0,89071	0,92066	0,96747	0,44177	1,55867	0,16604
0,42	0,88572	0,91697	0,96592	0,45218	1,52890	0,16927
0,43	0,88065	0,91322	0,96434	0,46257	1,50072	0,17245
0,44	0,87550	0,90940	0,96272	0,47293	1,47401	0,17558
0,45	0,87027	0,90551	0,96108	0,48326	1,44867	0,17865
0,46	0,86496	0,90157	0,95940	0,49357	1,42463	0,18166
0,47	0,85958	0,89756	0,95769	0,50385	1,40180	0,18462
0,48	0,85413	0,89349	0,95595	0,51410	1,38010	0,18753
0,49	0,84861	0,88936	0,95418	0,52433	1,35947	0,19037
0,50	0,84302	0,88517	0,95238	0,53452	1,33984	0,19316
0,51	0,83737	0,88093	0,95055	0,54469	1,32117	0,19589
0,52	0,83165	0,87663	0,94869	0,55483	1,30339	0,19856
0,53	0,82588	0,87228	0,94681	0,56493	1,28645	0,20118
0,54	0,82005	0,86788	0,94489	0,57501	1,27032	0,20373
0,55	0,81417	0,86342	0,94295	0,58506	1,25495	0,20623
0,56	0,80823	0,85892	0,94098	0,59507	1,24029	0,20866
0,57	0,80224	0,85437	0,93898	0,60505	1,22633	0,21104
0,58	0,79621	0,84978	0,93696	0,61501	1,21301	0,21336
0,59	0,79013	0,84514	0,93491	0,62492	1,20031	0,21561

0,60	0,78400	0,84045	0,93284	0,63481	1,18820	0,21781
0,61	0,77784	0,83573	0,93073	0,64466	1,17665	0,21995
0,62	0,77164	0,83096	0,92861	0,65448	1,16565	0,22203
0,63	0,76540	0,82616	0,92646	0,66427	1,15515	0,22404
0,64	0,75913	0,82132	0,92428	0,67402	1,14515	0,22600
0,65	0,75283	0,81644	0,92208	0,68374	1,13562	0,22790
0,66	0,74650	0,81153	0,91986	0,69342	1,12654	0,22973
0,67	0,74014	0,80659	0,91762	0,70307	1,11789	0,23151
0,68	0,73376	0,80162	0,91535	0,71268	1,10965	0,23323
0,69	0,72735	0,79661	0,91306	0,72225	1,10182	0,23489
0,70	0,72093	0,79158	0,91075	0,73179	1,09437	0,23649
0,71	0,71448	0,78652	0,90841	0,74129	1,08729	0,23803
0,72	0,70803	0,78143	0,90606	0,75076	1,08057	0,23951
0,73	0,70155	0,77632	0,90369	0,76019	1,07419	0,24093
0,74	0,69507	0,77119	0,90129	0,76958	1,06814	0,24229
0,75	0,68857	0,76604	0,89888	0,77894	1,06242	0,24360
0,76	0,68207	0,76086	0,89644	0,78825	1,05700	0,24485
0,77	0,67556	0,75567	0,89399	0,79753	1,05188	0,24604
0,78	0,66905	0,75046	0,89152	0,80677	1,04705	0,24717
0,79	0,66254	0,74523	0,88903	0,81597	1,04251	0,24825
0,80	0,65602	0,73999	0,88652	0,82514	1,03823	0,24927
0,81	0,64951	0,73474	0,88400	0,83426	1,03422	0,25024
0,82	0,64300	0,72947	0,88146	0,84335	1,03046	0,25115
0,83	0,63650	0,72419	0,87890	0,85239	1,02696	0,25201
0,84	0,63000	0,71891	0,87633	0,86140	1,02370	0,25281
0,85	0,62351	0,71361	0,87374	0,87037	1,02067	0,25356
0,86	0,61703	0,70831	0,87114	0,87929	1,01787	0,25426

0,87	0,61057	0,70300	0,86852	0,88818	1,01530	0,25490
0,88	0,60412	0,69768	0,86589	0,89703	1,01294	0,25550
0,89	0,59768	0,69236	0,86324	0,90583	1,01080	0,25604
0,90	0,59126	0,68704	0,86059	0,91460	1,00886	0,25653
0,91	0,58486	0,68172	0,85791	0,92332	1,00713	0,25697
0,92	0,57848	0,67640	0,85523	0,93201	1,00560	0,25736
0,93	0,57211	0,67108	0,85253	0,94065	1,00426	0,25771
0,94	0,56578	0,66576	0,84982	0,94925	1,00311	0,25800
0,95	0,55946	0,66044	0,84710	0,95781	1,00215	0,25825
0,96	0,55317	0,65513	0,84437	0,96633	1,00136	0,25845
0,97	0,54691	0,64982	0,84162	0,97481	1,00076	0,25861
0,98	0,54067	0,64452	0,83887	0,98325	1,00034	0,25872
0,99	0,53446	0,63923	0,83611	0,99165	1,00008	0,25878
1,00	0,52828	0,63394	0,83333	1,00000	1,00000	0,25880
1,01	0,52213	0,62866	0,83055	1,00831	1,00008	0,25878
1,02	0,51602	0,62339	0,82776	1,01658	1,00033	0,25872
1,03	0,50994	0,61813	0,82496	1,02481	1,00074	0,25861
1,04	0,50389	0,61289	0,82215	1,03300	1,00131	0,25847
1,05	0,49787	0,60765	0,81934	1,04114	1,00203	0,25828
1,06	0,49189	0,60243	0,81651	1,04925	1,00291	0,25805
1,07	0,48595	0,59722	0,81368	1,05731	1,00394	0,25779
1,08	0,48005	0,59203	0,81085	1,06533	1,00512	0,25749
1,09	0,47418	0,58686	0,80800	1,07331	1,00645	0,25715
1,10	0,46835	0,58170	0,80515	1,08124	1,00793	0,25677
1,11	0,46257	0,57655	0,80230	1,08913	1,00955	0,25636
1,12	0,45682	0,57143	0,79944	1,09699	1,01131	0,25591
1,13	0,45111	0,56632	0,79657	1,10479	1,01322	0,25543

1,14	0,44545	0,56123	0,79370	1,11256	1,01527	0,25491
1,15	0,43983	0,55616	0,79083	1,12029	1,01745	0,25436
1,16	0,43425	0,55112	0,78795	1,12797	1,01978	0,25378
1,17	0,42872	0,54609	0,78506	1,13561	1,02224	0,25317
1,18	0,42322	0,54108	0,78218	1,14321	1,02484	0,25253
1,19	0,41778	0,53610	0,77929	1,15077	1,02757	0,25186
1,20	0,41238	0,53114	0,77640	1,15828	1,03044	0,25116
1,21	0,40702	0,52620	0,77350	1,16575	1,03344	0,25043
1,22	0,40171	0,52129	0,77061	1,17319	1,03657	0,24967
1,23	0,39645	0,51640	0,76771	1,18057	1,03983	0,24889
1,24	0,39123	0,51154	0,76481	1,18792	1,04323	0,24808
1,25	0,38606	0,50670	0,76190	1,19523	1,04675	0,24724
1,26	0,38093	0,50189	0,75900	1,20249	1,05041	0,24638
1,27	0,37586	0,49710	0,75610	1,20972	1,05419	0,24550
1,28	0,37083	0,49234	0,75319	1,21690	1,05810	0,24459
1,29	0,36585	0,48761	0,75029	1,22404	1,06214	0,24366
1,30	0,36091	0,48290	0,74738	1,23114	1,06630	0,24271
1,31	0,35603	0,47822	0,74448	1,23819	1,07060	0,24174
1,32	0,35119	0,47357	0,74158	1,24521	1,07502	0,24074
1,33	0,34640	0,46895	0,73867	1,25218	1,07957	0,23973
1,34	0,34166	0,46436	0,73577	1,25912	1,08424	0,23870
1,35	0,33697	0,45980	0,73287	1,26601	1,08904	0,23764
1,36	0,33233	0,45526	0,72997	1,27286	1,09396	0,23657
1,37	0,32773	0,45076	0,72707	1,27968	1,09902	0,23549
1,38	0,32319	0,44628	0,72418	1,28645	1,10419	0,23438
1,39	0,31869	0,44184	0,72128	1,29318	1,10950	0,23326
1,40	0,31424	0,43742	0,71839	1,29987	1,11493	0,23213

1,41	0,30984	0,43304	0,71550	1,30652	1,12048	0,23098
1,42	0,30549	0,42869	0,71262	1,31313	1,12616	0,22981
1,43	0,30118	0,42436	0,70973	1,31970	1,13197	0,22863
1,44	0,29693	0,42007	0,70685	1,32623	1,13790	0,22744
1,45	0,29272	0,41581	0,70398	1,33272	1,14396	0,22623
1,46	0,28856	0,41158	0,70110	1,33917	1,15015	0,22502
1,47	0,28445	0,40739	0,69824	1,34558	1,15646	0,22379
1,48	0,28039	0,40322	0,69537	1,35195	1,16290	0,22255
1,49	0,27637	0,39909	0,69251	1,35828	1,16947	0,22130
1,50	0,27240	0,39498	0,68966	1,36458	1,17617	0,22004
1,51	0,26848	0,39091	0,68680	1,37083	1,18299	0,21877
1,52	0,26461	0,38688	0,68396	1,37705	1,18994	0,21749
1,53	0,26078	0,38287	0,68112	1,38322	1,19702	0,21621
1,54	0,25700	0,37890	0,67828	1,38936	1,20423	0,21491
1,55	0,25326	0,37495	0,67545	1,39546	1,21157	0,21361
1,56	0,24957	0,37105	0,67262	1,40152	1,21904	0,21230
1,57	0,24593	0,36717	0,66980	1,40755	1,22664	0,21099
1,58	0,24233	0,36332	0,66699	1,41353	1,23438	0,20966
1,59	0,23878	0,35951	0,66418	1,41948	1,24224	0,20834
1,60	0,23527	0,35573	0,66138	1,42539	1,25024	0,20700
1,61	0,23181	0,35198	0,65858	1,43127	1,25836	0,20567
1,62	0,22839	0,34827	0,65579	1,43710	1,26663	0,20433
1,63	0,22501	0,34458	0,65301	1,44290	1,27502	0,20298
1,64	0,22168	0,34093	0,65023	1,44866	1,28355	0,20163
1,65	0,21839	0,33731	0,64746	1,45439	1,29222	0,20028
1,66	0,21515	0,33372	0,64470	1,46008	1,30102	0,19892
1,67	0,21195	0,33017	0,64194	1,46573	1,30996	0,19757

1,68	0,20879	0,32664	0,63919	1,47135	1,31904	0,19621
1,69	0,20567	0,32315	0,63645	1,47693	1,32825	0,19485
1,70	0,20259	0,31969	0,63371	1,48247	1,33761	0,19348
1,71	0,19956	0,31626	0,63099	1,48798	1,34710	0,19212
1,72	0,19656	0,31287	0,62827	1,49345	1,35674	0,19076
1,73	0,19361	0,30950	0,62556	1,49889	1,36651	0,18939
1,74	0,19070	0,30617	0,62285	1,50429	1,37643	0,18803
1,75	0,18782	0,30287	0,62016	1,50966	1,38649	0,18666
1,76	0,18499	0,29959	0,61747	1,51499	1,39670	0,18530
1,77	0,18219	0,29635	0,61479	1,52029	1,40705	0,18393
1,78	0,17944	0,29315	0,61211	1,52555	1,41755	0,18257
1,79	0,17672	0,28997	0,60945	1,53078	1,42819	0,18121
1,80	0,17404	0,28682	0,60680	1,53598	1,43898	0,17985
1,81	0,17140	0,28370	0,60415	1,54114	1,44992	0,17850
1,82	0,16879	0,28061	0,60151	1,54626	1,46101	0,17714
1,83	0,16622	0,27756	0,59888	1,55136	1,47225	0,17579
1,84	0,16369	0,27453	0,59626	1,55642	1,48365	0,17444
1,85	0,16119	0,27153	0,59365	1,56145	1,49519	0,17309
1,86	0,15873	0,26857	0,59104	1,56644	1,50689	0,17175
1,87	0,15631	0,26563	0,58845	1,57140	1,51875	0,17041
1,88	0,15392	0,26272	0,58586	1,57633	1,53076	0,16907
1,89	0,15156	0,25984	0,58329	1,58123	1,54293	0,16774
1,90	0,14924	0,25699	0,58072	1,58609	1,55526	0,16641
1,91	0,14695	0,25417	0,57816	1,59092	1,56774	0,16508
1,92	0,14470	0,25138	0,57561	1,59572	1,58039	0,16376
1,93	0,14247	0,24861	0,57307	1,60049	1,59320	0,16244
1,94	0,14028	0,24588	0,57054	1,60523	1,60617	0,16113

1,95	0,13813	0,24317	0,56802	1,60993	1,61931	0,15982
1,96	0,13600	0,24049	0,56551	1,61460	1,63261	0,15852
1,97	0,13390	0,23784	0,56301	1,61925	1,64608	0,15722
1,98	0,13184	0,23521	0,56051	1,62386	1,65972	0,15593
1,99	0,12981	0,23262	0,55803	1,62844	1,67352	0,15465
2,00	0,12780	0,23005	0,55556	1,63299	1,68750	0,15337
2,01	0,12583	0,22751	0,55309	1,63751	1,70165	0,15209
2,02	0,12389	0,22499	0,55064	1,64201	1,71597	0,15082
2,03	0,12197	0,22250	0,54819	1,64647	1,73047	0,14956
2,04	0,12009	0,22004	0,54576	1,65090	1,74514	0,14830
2,05	0,11823	0,21760	0,54333	1,65530	1,75999	0,14705
2,06	0,11640	0,21519	0,54091	1,65967	1,77502	0,14580
2,07	0,11460	0,21281	0,53851	1,66402	1,79022	0,14457
2,08	0,11282	0,21045	0,53611	1,66833	1,80561	0,14333
2,09	0,11107	0,20811	0,53373	1,67262	1,82119	0,14211
2,10	0,10935	0,20580	0,53135	1,67687	1,83694	0,14089
2,11	0,10766	0,20352	0,52898	1,68110	1,85289	0,13968
2,12	0,10599	0,20126	0,52663	1,68530	1,86902	0,13847
2,13	0,10434	0,19902	0,52428	1,68947	1,88533	0,13727
2,14	0,10273	0,19681	0,52194	1,69362	1,90184	0,13608
2,15	0,10113	0,19463	0,51962	1,69774	1,91854	0,13490
2,16	0,09956	0,19247	0,51730	1,70183	1,93544	0,13372
2,17	0,09802	0,19033	0,51499	1,70589	1,95252	0,13255
2,18	0,09649	0,18821	0,51269	1,70992	1,96981	0,13139
2,19	0,09500	0,18612	0,51041	1,71393	1,98729	0,13023
2,20	0,09352	0,18405	0,50813	1,71791	2,00497	0,12908
2,21	0,09207	0,18200	0,50586	1,72187	2,02286	0,12794

2,22	0,09064	0,17998	0,50361	1,72579	2,04094	0,12681
2,23	0,08923	0,17798	0,50136	1,72970	2,05923	0,12568
2,24	0,08785	0,17600	0,49912	1,73357	2,07773	0,12456
2,25	0,08648	0,17404	0,49689	1,73742	2,09644	0,12345
2,26	0,08514	0,17211	0,49468	1,74125	2,11535	0,12235
2,27	0,08382	0,17020	0,49247	1,74504	2,13447	0,12125
2,28	0,08251	0,16830	0,49027	1,74882	2,15381	0,12016
2,29	0,08123	0,16643	0,48809	1,75257	2,17336	0,11908
2,30	0,07997	0,16458	0,48591	1,75629	2,19313	0,11801
2,31	0,07873	0,16275	0,48374	1,75999	2,21312	0,11694
2,32	0,07751	0,16095	0,48158	1,76366	2,23332	0,11588
2,33	0,07631	0,15916	0,47944	1,76731	2,25375	0,11483
2,34	0,07512	0,15739	0,47730	1,77093	2,27440	0,11379
2,35	0,07396	0,15564	0,47517	1,77453	2,29528	0,11276
2,36	0,07281	0,15391	0,47305	1,77811	2,31638	0,11173
2,37	0,07168	0,15221	0,47095	1,78166	2,33771	0,11071
2,38	0,07057	0,15052	0,46885	1,78519	2,35928	0,10970
2,39	0,06948	0,14885	0,46676	1,78869	2,38107	0,10869
2,40	0,06840	0,14720	0,46468	1,79218	2,40310	0,10770
2,41	0,06734	0,14556	0,46262	1,79563	2,42537	0,10671
2,42	0,06630	0,14395	0,46056	1,79907	2,44787	0,10573
2,43	0,06527	0,14235	0,45851	1,80248	2,47061	0,10475
2,44	0,06426	0,14078	0,45647	1,80587	2,49360	0,10379
2,45	0,06327	0,13922	0,45444	1,80924	2,51683	0,10283
2,46	0,06229	0,13768	0,45242	1,81258	2,54031	0,10188
2,47	0,06133	0,13615	0,45041	1,81591	2,56403	0,10094
2,48	0,06038	0,13465	0,44841	1,81921	2,58801	0,10000

2,49	0,05945	0,13316	0,44642	1,82249	2,61224	0,09907
2,50	0,05853	0,13169	0,44444	1,82574	2,63672	0,09815
2,51	0,05762	0,13023	0,44247	1,82898	2,66146	0,09724
2,52	0,05674	0,12879	0,44051	1,83219	2,68645	0,09634
2,53	0,05586	0,12737	0,43856	1,83538	2,71171	0,09544
2,54	0,05500	0,12597	0,43662	1,83855	2,73723	0,09455
2,55	0,05415	0,12458	0,43469	1,84170	2,76301	0,09367
2,56	0,05332	0,12321	0,43277	1,84483	2,78906	0,09279
2,57	0,05250	0,12185	0,43085	1,84794	2,81538	0,09193
2,58	0,05169	0,12051	0,42895	1,85103	2,84197	0,09106
2,59	0,05090	0,11918	0,42705	1,85410	2,86884	0,09021
2,60	0,05012	0,11787	0,42517	1,85714	2,89598	0,08937
2,61	0,04935	0,11658	0,42329	1,86017	2,92339	0,08853
2,62	0,04859	0,11530	0,42143	1,86318	2,95109	0,08770
2,63	0,04784	0,11403	0,41957	1,86616	2,97907	0,08687
2,64	0,04711	0,11278	0,41772	1,86913	3,00733	0,08606
2,65	0,04639	0,11154	0,41589	1,87208	3,03588	0,08525
2,66	0,04568	0,11032	0,41406	1,87501	3,06472	0,08445
2,67	0,04498	0,10911	0,41224	1,87792	3,09385	0,08365
2,68	0,04429	0,10792	0,41043	1,88081	3,12327	0,08286
2,69	0,04362	0,10674	0,40863	1,88368	3,15299	0,08208
2,70	0,04295	0,10557	0,40683	1,88653	3,18301	0,08131
2,71	0,04229	0,10442	0,40505	1,88936	3,21333	0,08054
2,72	0,04165	0,10328	0,40328	1,89218	3,24395	0,07978
2,73	0,04102	0,10215	0,40151	1,89497	3,27488	0,07903
2,74	0,04039	0,10104	0,39976	1,89775	3,30611	0,07828
2,75	0,03978	0,09994	0,39801	1,90051	3,33766	0,07754

2,76	0,03917	0,09885	0,39627	1,90325	3,36952	0,07681
2,77	0,03858	0,09778	0,39454	1,90598	3,40169	0,07608
2,78	0,03799	0,09671	0,39282	1,90868	3,43418	0,07536
2,79	0,03742	0,09566	0,39111	1,91137	3,46699	0,07465
2,80	0,03685	0,09463	0,38941	1,91404	3,50012	0,07394
2,81	0,03629	0,09360	0,38771	1,91669	3,53358	0,07324
2,82	0,03574	0,09259	0,38603	1,91933	3,56737	0,07255
2,83	0,03520	0,09158	0,38435	1,92195	3,60148	0,07186
2,84	0,03467	0,09059	0,38268	1,92455	3,63593	0,07118
2,85	0,03415	0,08962	0,38102	1,92714	3,67072	0,07051
2,86	0,03363	0,08865	0,37937	1,92970	3,70584	0,06984
2,87	0,03312	0,08769	0,37773	1,93225	3,74131	0,06917
2,88	0,03263	0,08675	0,37610	1,93479	3,77711	0,06852
2,89	0,03213	0,08581	0,37447	1,93731	3,81327	0,06787
2,90	0,03165	0,08489	0,37286	1,93981	3,84977	0,06723
2,91	0,03118	0,08398	0,37125	1,94230	3,88662	0,06659
2,92	0,03071	0,08307	0,36965	1,94477	3,92383	0,06596
2,93	0,03025	0,08218	0,36806	1,94722	3,96139	0,06533
2,94	0,02980	0,08130	0,36647	1,94966	3,99932	0,06471
2,95	0,02935	0,08043	0,36490	1,95208	4,03760	0,06410
2,96	0,02891	0,07957	0,36333	1,95449	4,07625	0,06349
2,97	0,02848	0,07872	0,36177	1,95688	4,11527	0,06289
2,98	0,02805	0,07788	0,36022	1,95925	4,15466	0,06229
2,99	0,02764	0,07705	0,35868	1,96162	4,19443	0,06170
3,00	0,02722	0,07623	0,35714	1,96396	4,23457	0,06112
3,01	0,02682	0,07541	0,35562	1,96629	4,27509	0,06054
3,02	0,02642	0,07461	0,35410	1,96861	4,31599	0,05996

3,03	0,02603	0,07382	0,35259	1,97091	4,35728	0,05940
3,04	0,02564	0,07303	0,35108	1,97319	4,39895	0,05883
3,05	0,02526	0,07226	0,34959	1,97547	4,44102	0,05828
3,06	0,02489	0,07149	0,34810	1,97772	4,48347	0,05772
3,07	0,02452	0,07074	0,34662	1,97997	4,52633	0,05718
3,08	0,02416	0,06999	0,34515	1,98219	4,56959	0,05664
3,09	0,02380	0,06925	0,34369	1,98441	4,61325	0,05610
3,10	0,02345	0,06852	0,34223	1,98661	4,65731	0,05557
3,11	0,02310	0,06779	0,34078	1,98879	4,70178	0,05504
3,12	0,02276	0,06708	0,33934	1,99097	4,74667	0,05452
3,13	0,02243	0,06637	0,33791	1,99313	4,79197	0,05401
3,14	0,02210	0,06568	0,33648	1,99527	4,83769	0,05350
3,15	0,02177	0,06499	0,33506	1,99740	4,88383	0,05299
3,16	0,02146	0,06430	0,33365	1,99952	4,93039	0,05249
3,17	0,02114	0,06363	0,33225	2,00162	4,97739	0,05200
3,18	0,02083	0,06296	0,33085	2,00372	5,02481	0,05151
3,19	0,02053	0,06231	0,32947	2,00579	5,07266	0,05102
3,20	0,02023	0,06165	0,32808	2,00786	5,12096	0,05054
3,21	0,01993	0,06101	0,32671	2,00991	5,16969	0,05006
3,22	0,01964	0,06037	0,32534	2,01195	5,21887	0,04959
3,23	0,01936	0,05975	0,32398	2,01398	5,26849	0,04912
3,24	0,01908	0,05912	0,32263	2,01599	5,31857	0,04866
3,25	0,01880	0,05851	0,32129	2,01799	5,36909	0,04820
3,26	0,01853	0,05790	0,31995	2,01998	5,42008	0,04775
3,27	0,01826	0,05730	0,31862	2,02196	5,47152	0,04730
3,28	0,01799	0,05671	0,31729	2,02392	5,52343	0,04686
3,29	0,01773	0,05612	0,31597	2,02587	5,57580	0,04642

3,30	0,01748	0,05554	0,31466	2,02781	5,62865	0,04598
3,31	0,01722	0,05497	0,31336	2,02974	5,68196	0,04555
3,32	0,01698	0,05440	0,31206	2,03165	5,73576	0,04512
3,33	0,01673	0,05384	0,31077	2,03356	5,79003	0,04470
3,34	0,01649	0,05329	0,30949	2,03545	5,84479	0,04428
3,35	0,01625	0,05274	0,30821	2,03733	5,90004	0,04386
3,36	0,01602	0,05220	0,30694	2,03920	5,95577	0,04345
3,37	0,01579	0,05166	0,30568	2,04106	6,01201	0,04305
3,38	0,01557	0,05113	0,30443	2,04290	6,06873	0,04265
3,39	0,01534	0,05061	0,30318	2,04474	6,12596	0,04225
3,40	0,01512	0,05009	0,30193	2,04656	6,18370	0,04185
3,41	0,01491	0,04958	0,30070	2,04837	6,24194	0,04146
3,42	0,01470	0,04908	0,29947	2,05017	6,30070	0,04108
3,43	0,01449	0,04858	0,29824	2,05196	6,35997	0,04069
3,44	0,01428	0,04808	0,29702	2,05374	6,41976	0,04031
3,45	0,01408	0,04759	0,29581	2,05551	6,48007	0,03994
3,46	0,01388	0,04711	0,29461	2,05727	6,54092	0,03957
3,47	0,01368	0,04663	0,29341	2,05901	6,60229	0,03920
3,48	0,01349	0,04616	0,29222	2,06075	6,66419	0,03884
3,49	0,01330	0,04569	0,29103	2,06247	6,72664	0,03847
3,50	0,01311	0,04523	0,28986	2,06419	6,78962	0,03812
3,51	0,01293	0,04478	0,28868	2,06589	6,85315	0,03776
3,52	0,01274	0,04433	0,28751	2,06759	6,91723	0,03741
3,53	0,01256	0,04388	0,28635	2,06927	6,98186	0,03707
3,54	0,01239	0,04344	0,28520	2,07094	7,04705	0,03673
3,55	0,01221	0,04300	0,28405	2,07261	7,11281	0,03639
3,56	0,01204	0,04257	0,28291	2,07426	7,17912	0,03605

3,57	0,01188	0,04214	0,28177	2,07590	7,24601	0,03572
3,58	0,01171	0,04172	0,28064	2,07754	7,31346	0,03539
3,59	0,01155	0,04131	0,27952	2,07916	7,38150	0,03506
3,60	0,01138	0,04089	0,27840	2,08077	7,45011	0,03474
3,61	0,01123	0,04049	0,27728	2,08238	7,51931	0,03442
3,62	0,01107	0,04008	0,27618	2,08397	7,58910	0,03410
3,63	0,01092	0,03968	0,27507	2,08556	7,65948	0,03379
3,64	0,01076	0,03929	0,27398	2,08713	7,73045	0,03348
3,65	0,01062	0,03890	0,27289	2,08870	7,80203	0,03317
3,66	0,01047	0,03852	0,27180	2,09026	7,87421	0,03287
3,67	0,01032	0,03813	0,27073	2,09180	7,94700	0,03257
3,68	0,01018	0,03776	0,26965	2,09334	8,02040	0,03227
3,69	0,01004	0,03739	0,26858	2,09487	8,09442	0,03197
3,70	0,00990	0,03702	0,26752	2,09639	8,16907	0,03168
3,71	0,00977	0,03665	0,26647	2,09790	8,24433	0,03139
3,72	0,00963	0,03629	0,26542	2,09941	8,32023	0,03111
3,73	0,00950	0,03594	0,26437	2,10090	8,39676	0,03082
3,74	0,00937	0,03558	0,26333	2,10238	8,47393	0,03054
3,75	0,00924	0,03524	0,26230	2,10386	8,55174	0,03026
3,76	0,00912	0,03489	0,26127	2,10533	8,63020	0,02999
3,77	0,00899	0,03455	0,26024	2,10679	8,70931	0,02972
3,78	0,00887	0,03421	0,25922	2,10824	8,78907	0,02945
3,79	0,00875	0,03388	0,25821	2,10968	8,86950	0,02918
3,80	0,00863	0,03355	0,25720	2,11111	8,95059	0,02891
3,81	0,00851	0,03322	0,25620	2,11254	9,03234	0,02865
3,82	0,00840	0,03290	0,25520	2,11395	9,11477	0,02839
3,83	0,00828	0,03258	0,25421	2,11536	9,19788	0,02814

3,84	0,00817	0,03227	0,25322	2,11676	9,28167	0,02788
3,85	0,00806	0,03195	0,25224	2,11815	9,36614	0,02763
3,86	0,00795	0,03165	0,25126	2,11954	9,45131	0,02738
3,87	0,00784	0,03134	0,25029	2,12091	9,53717	0,02714
3,88	0,00774	0,03104	0,24932	2,12228	9,62373	0,02689
3,89	0,00763	0,03074	0,24836	2,12364	9,71100	0,02665
3,90	0,00753	0,03044	0,24740	2,12499	9,79897	0,02641
3,91	0,00743	0,03015	0,24645	2,12634	9,88766	0,02617
3,92	0,00733	0,02986	0,24550	2,12767	9,97707	0,02594
3,93	0,00723	0,02958	0,24456	2,12900	10,06720	0,02571
3,94	0,00714	0,02929	0,24362	2,13032	10,15806	0,02548
3,95	0,00704	0,02902	0,24269	2,13163	10,24965	0,02525
3,96	0,00695	0,02874	0,24176	2,13294	10,34197	0,02502
3,97	0,00686	0,02846	0,24084	2,13424	10,43504	0,02480
3,98	0,00676	0,02819	0,23992	2,13553	10,52886	0,02458
3,99	0,00667	0,02793	0,23900	2,13681	10,62343	0,02436
4,00	0,00659	0,02766	0,23810	2,13809	10,71875	0,02414
4,01	0,00650	0,02740	0,23719	2,13936	10,81483	0,02393
4,02	0,00641	0,02714	0,23629	2,14062	10,91168	0,02372
4,03	0,00633	0,02688	0,23539	2,14188	11,00931	0,02351
4,04	0,00624	0,02663	0,23450	2,14312	11,10770	0,02330
4,05	0,00616	0,02638	0,23362	2,14436	11,20688	0,02309
4,06	0,00608	0,02613	0,23274	2,14560	11,30684	0,02289
4,07	0,00600	0,02589	0,23186	2,14682	11,40760	0,02269
4,08	0,00592	0,02564	0,23099	2,14804	11,50915	0,02249
4,09	0,00585	0,02540	0,23012	2,14926	11,61150	0,02229
4,10	0,00577	0,02516	0,22925	2,15046	11,71465	0,02209

4,11	0,00569	0,02493	0,22839	2,15166	11,81862	0,02190
4,12	0,00562	0,02470	0,22754	2,15285	11,92340	0,02171
4,13	0,00555	0,02447	0,22669	2,15404	12,02900	0,02152
4,14	0,00547	0,02424	0,22584	2,15522	12,13543	0,02133
4,15	0,00540	0,02401	0,22500	2,15639	12,24269	0,02114
4,16	0,00533	0,02379	0,22416	2,15756	12,35079	0,02095
4,17	0,00526	0,02357	0,22332	2,15871	12,45973	0,02077
4,18	0,00520	0,02335	0,22250	2,15987	12,56951	0,02059
4,19	0,00513	0,02313	0,22167	2,16101	12,68015	0,02041
4,20	0,00506	0,02292	0,22085	2,16215	12,79164	0,02023
4,21	0,00500	0,02271	0,22003	2,16329	12,90400	0,02006
4,22	0,00493	0,02250	0,21922	2,16442	13,01722	0,01988
4,23	0,00487	0,02229	0,21841	2,16554	13,13132	0,01971
4,24	0,00481	0,02209	0,21760	2,16665	13,24629	0,01954
4,25	0,00474	0,02189	0,21680	2,16776	13,36215	0,01937
4,26	0,00468	0,02169	0,21601	2,16886	13,47890	0,01920
4,27	0,00462	0,02149	0,21521	2,16996	13,59655	0,01903
4,28	0,00457	0,02129	0,21442	2,17105	13,71509	0,01887
4,29	0,00451	0,02110	0,21364	2,17214	13,83454	0,01871
4,30	0,00445	0,02090	0,21286	2,17321	13,95490	0,01855
4,31	0,00439	0,02071	0,21208	2,17429	14,07618	0,01839
4,32	0,00434	0,02052	0,21131	2,17535	14,19838	0,01823
4,33	0,00428	0,02034	0,21054	2,17642	14,32151	0,01807
4,34	0,00423	0,02015	0,20977	2,17747	14,44557	0,01792
4,35	0,00417	0,01997	0,20901	2,17852	14,57057	0,01776
4,36	0,00412	0,01979	0,20825	2,17956	14,69652	0,01761
4,37	0,00407	0,01961	0,20750	2,18060	14,82342	0,01746

4,38	0,00402	0,01944	0,20674	2,18163	14,95127	0,01731
4,39	0,00397	0,01926	0,20600	2,18266	15,08008	0,01716
4,40	0,00392	0,01909	0,20525	2,18368	15,20987	0,01702
4,41	0,00387	0,01892	0,20451	2,18470	15,34062	0,01687
4,42	0,00382	0,01875	0,20378	2,18571	15,47236	0,01673
4,43	0,00377	0,01858	0,20305	2,18671	15,60508	0,01658
4,44	0,00372	0,01841	0,20232	2,18771	15,73879	0,01644
4,45	0,00368	0,01825	0,20159	2,18871	15,87350	0,01630
4,46	0,00363	0,01808	0,20087	2,18970	16,00921	0,01617
4,47	0,00359	0,01792	0,20015	2,19068	16,14593	0,01603
4,48	0,00354	0,01776	0,19944	2,19166	16,28366	0,01589
4,49	0,00350	0,01761	0,19873	2,19263	16,42241	0,01576
4,50	0,00346	0,01745	0,19802	2,19360	16,56219	0,01563
4,51	0,00341	0,01729	0,19732	2,19456	16,70301	0,01549
4,52	0,00337	0,01714	0,19662	2,19552	16,84486	0,01536
4,53	0,00333	0,01699	0,19592	2,19647	16,98776	0,01523
4,54	0,00329	0,01684	0,19522	2,19742	17,13170	0,01511
4,55	0,00325	0,01669	0,19453	2,19836	17,27671	0,01498
4,56	0,00321	0,01654	0,19385	2,19930	17,42277	0,01485
4,57	0,00317	0,01640	0,19316	2,20023	17,56991	0,01473
4,58	0,00313	0,01625	0,19248	2,20116	17,71812	0,01461
4,59	0,00309	0,01611	0,19181	2,20208	17,86741	0,01448
4,60	0,00305	0,01597	0,19113	2,20300	18,01779	0,01436
4,61	0,00302	0,01583	0,19046	2,20391	18,16927	0,01424
4,62	0,00298	0,01569	0,18979	2,20482	18,32185	0,01413
4,63	0,00294	0,01556	0,18913	2,20572	18,47553	0,01401
4,64	0,00291	0,01542	0,18847	2,20662	18,63032	0,01389

4,65	0,00287	0,01529	0,18781	2,20752	18,78624	0,01378
4,66	0,00284	0,01515	0,18716	2,20841	18,94328	0,01366
4,67	0,00280	0,01502	0,18651	2,20929	19,10145	0,01355
4,68	0,00277	0,01489	0,18586	2,21017	19,26076	0,01344
4,69	0,00273	0,01476	0,18521	2,21105	19,42122	0,01333
4,70	0,00270	0,01464	0,18457	2,21192	19,58283	0,01322
4,71	0,00267	0,01451	0,18393	2,21278	19,74559	0,01311
4,72	0,00264	0,01438	0,18330	2,21365	19,90953	0,01300
4,73	0,00260	0,01426	0,18266	2,21450	20,07463	0,01289
4,74	0,00257	0,01414	0,18203	2,21536	20,24091	0,01279
4,75	0,00254	0,01402	0,18141	2,21621	20,40838	0,01268
4,76	0,00251	0,01390	0,18078	2,21705	20,57703	0,01258
4,77	0,00248	0,01378	0,18016	2,21789	20,74689	0,01247
4,78	0,00245	0,01366	0,17954	2,21872	20,91795	0,01237
4,79	0,00242	0,01354	0,17893	2,21956	21,09022	0,01227
4,80	0,00239	0,01343	0,17832	2,22038	21,26371	0,01217
4,81	0,00237	0,01331	0,17771	2,22121	21,43842	0,01207
4,82	0,00234	0,01320	0,17710	2,22202	21,61437	0,01197
4,83	0,00231	0,01309	0,17650	2,22284	21,79156	0,01188
4,84	0,00228	0,01298	0,17590	2,22365	21,96999	0,01178
4,85	0,00226	0,01287	0,17530	2,22445	22,14967	0,01168
4,86	0,00223	0,01276	0,17471	2,22526	22,33061	0,01159
4,87	0,00220	0,01265	0,17411	2,22605	22,51282	0,01150
4,88	0,00218	0,01254	0,17352	2,22685	22,69631	0,01140
4,89	0,00215	0,01244	0,17294	2,22764	22,88107	0,01131
4,90	0,00213	0,01233	0,17235	2,22842	23,06712	0,01122
4,91	0,00210	0,01223	0,17177	2,22921	23,25446	0,01113

4,92	0,00208	0,01213	0,17120	2,22998	23,44311	0,01104
4,93	0,00205	0,01202	0,17062	2,23076	23,63307	0,01095
4,94	0,00203	0,01192	0,17005	2,23153	23,82434	0,01086
4,95	0,00200	0,01182	0,16948	2,23229	24,01693	0,01078
4,96	0,00198	0,01173	0,16891	2,23306	24,21086	0,01069
4,97	0,00196	0,01163	0,16835	2,23382	24,40612	0,01060
4,98	0,00193	0,01153	0,16778	2,23457	24,60272	0,01052
4,99	0,00191	0,01144	0,16722	2,23532	24,80068	0,01044
5,00	0,00189	0,01134	0,16667	2,23607	25,00000	0,01035

A.5. Straight shock table

The treatment of the previous examples resorted to a considerable number of formulae. As already stressed, these formulae result from a limited number of physical laws and constraints.

In the treatment of the posed problems, we are free to choose at each stage the formula corresponding to the step of the adopted reasoning. This requires some form of programming.

Similarly to the previous chapter, one can also definitively tabulate all the results yielded by formulae of practical interest.

The tables presented here have been very easily built with a classical spreadsheet. These tables depend on the value of the polytropic exponent $\gamma = \frac{c_P}{c_V}$.

The following tables have been set up for $\gamma = 1.4$, the "standard" value that is valid for the air and several other gases to a satisfactory degree of approximation. Similarly to the previous chapter, all the problems are reported to the use of this value of γ.

The programming stage cannot be avoided in case of repetitive computations, but it can readily integrate $\gamma = \frac{c_P}{c_V}$ as a parameter. Having built the table for one parameter, therefore for one fluid, the same programmed table can be directly reused for another fluid.

Below are the formulae to be programmed for each column of the table:

Straight shock table	
M_2	Equation [3.8]
p_2 / p_1	Equation [3.5]
ρ_2 / ρ_1	Equation [3.6]
T_2 / T_1	Equation [3.7]
p_{01} / p_{02}	Equation [3.12]

Straight shock tables. $\gamma = 1{,}4$:

M_1	M_2	p_2/p_1	ρ_2/ρ_1	T_2/T_1	p_{01}/p_{02}
1,0000	1,0000	1,0000	1,0000	1,0000	1,0000
1,0100	0,9901	1,0235	1,0167	1,0066	1,0000
1,0200	0,9805	1,0471	1,0334	1,0132	1,0000
1,0300	0,9712	1,0711	1,0502	1,0198	1,0000
1,0400	0,9620	1,0952	1,0671	1,0263	0,9999
1,0500	0,9531	1,1196	1,0840	1,0328	0,9999
1,0600	0,9444	1,1442	1,1009	1,0393	0,9998
1,0700	0,9360	1,1691	1,1179	1,0458	0,9996
1,0800	0,9277	1,1941	1,1349	1,0522	0,9994
1,0900	0,9196	1,2195	1,1520	1,0586	0,9992
1,1000	0,9118	1,2450	1,1691	1,0649	0,9989
1,1100	0,9041	1,2708	1,1862	1,0713	0,9986
1,1200	0,8966	1,2968	1,2034	1,0776	0,9982
1,1300	0,8892	1,3231	1,2206	1,0840	0,9978
1,1400	0,8820	1,3495	1,2378	1,0903	0,9973

1,1500	0,8750	1,3763	1,2550	1,0966	0,9967
1,1600	0,8682	1,4032	1,2723	1,1029	0,9961
1,1700	0,8615	1,4304	1,2896	1,1092	0,9953
1,1800	0,8549	1,4578	1,3069	1,1154	0,9946
1,1900	0,8485	1,4855	1,3243	1,1217	0,9937
1,2000	0,8422	1,5133	1,3416	1,1280	0,9928
1,2100	0,8360	1,5415	1,3590	1,1343	0,9918
1,2200	0,8300	1,5698	1,3764	1,1405	0,9907
1,2300	0,8241	1,5984	1,3938	1,1468	0,9896
1,2400	0,8183	1,6272	1,4112	1,1531	0,9884
1,2500	0,8126	1,6563	1,4286	1,1594	0,9871
1,2600	0,8071	1,6855	1,4460	1,1657	0,9857
1,2700	0,8016	1,7151	1,4634	1,1720	0,9842
1,2800	0,7963	1,7448	1,4808	1,1783	0,9827
1,2900	0,7911	1,7748	1,4983	1,1846	0,9811
1,3000	0,7860	1,8050	1,5157	1,1909	0,9794
1,3100	0,7809	1,8355	1,5331	1,1972	0,9776
1,3200	0,7760	1,8661	1,5505	1,2035	0,9758
1,3300	0,7712	1,8971	1,5680	1,2099	0,9738
1,3400	0,7664	1,9282	1,5854	1,2162	0,9718
1,3500	0,7618	1,9596	1,6028	1,2226	0,9697
1,3600	0,7572	1,9912	1,6202	1,2290	0,9676
1,3700	0,7527	2,0231	1,6376	1,2354	0,9653
1,3800	0,7483	2,0551	1,6549	1,2418	0,9630
1,3900	0,7440	2,0875	1,6723	1,2482	0,9607
1,4000	0,7397	2,1200	1,6897	1,2547	0,9582
1,4100	0,7355	2,1528	1,7070	1,2612	0,9557
1,4200	0,7314	2,1858	1,7243	1,2676	0,9531
1,4300	0,7274	2,2191	1,7416	1,2741	0,9504
1,4400	0,7235	2,2525	1,7589	1,2807	0,9476
1,4500	0,7196	2,2863	1,7761	1,2872	0,9448
1,4600	0,7157	2,3202	1,7934	1,2938	0,9420

1,4700	0,7120	2,3544	1,8106	1,3003	0,9390
1,4800	0,7083	2,3888	1,8278	1,3069	0,9360
1,4900	0,7047	2,4235	1,8449	1,3136	0,9329
1,5000	0,7011	2,4583	1,8621	1,3202	0,9298
1,5100	0,6976	2,4935	1,8792	1,3269	0,9266
1,5200	0,6941	2,5288	1,8963	1,3336	0,9233
1,5300	0,6907	2,5644	1,9133	1,3403	0,9200
1,5400	0,6874	2,6002	1,9303	1,3470	0,9166
1,5500	0,6841	2,6363	1,9473	1,3538	0,9132
1,5600	0,6809	2,6725	1,9643	1,3606	0,9097
1,5700	0,6777	2,7091	1,9812	1,3674	0,9062
1,5800	0,6746	2,7458	1,9981	1,3742	0,9026
1,5900	0,6715	2,7828	2,0149	1,3811	0,8989
1,6000	0,6684	2,8200	2,0317	1,3880	0,8952
1,6100	0,6655	2,8575	2,0485	1,3949	0,8915
1,6200	0,6625	2,8951	2,0653	1,4018	0,8877
1,6300	0,6596	2,9331	2,0820	1,4088	0,8838
1,6400	0,6568	2,9712	2,0986	1,4158	0,8799
1,6500	0,6540	3,0096	2,1152	1,4228	0,8760
1,6600	0,6512	3,0482	2,1318	1,4299	0,8720
1,6700	0,6485	3,0871	2,1484	1,4369	0,8680
1,6800	0,6458	3,1261	2,1649	1,4440	0,8639
1,6900	0,6431	3,1655	2,1813	1,4512	0,8599
1,7000	0,6405	3,2050	2,1977	1,4583	0,8557
1,7100	0,6380	3,2448	2,2141	1,4655	0,8516
1,7200	0,6355	3,2848	2,2304	1,4727	0,8474
1,7300	0,6330	3,3251	2,2467	1,4800	0,8431
1,7400	0,6305	3,3655	2,2629	1,4873	0,8389
1,7500	0,6281	3,4063	2,2791	1,4946	0,8346
1,7600	0,6257	3,4472	2,2952	1,5019	0,8302
1,7700	0,6234	3,4884	2,3113	1,5093	0,8259
1,7800	0,6210	3,5298	2,3273	1,5167	0,8215

1,7900	0,6188	3,5715	2,3433	1,5241	0,8171
1,8000	0,6165	3,6133	2,3592	1,5316	0,8127
1,8100	0,6143	3,6555	2,3751	1,5391	0,8082
1,8200	0,6121	3,6978	2,3909	1,5466	0,8038
1,8300	0,6099	3,7404	2,4067	1,5541	0,7993
1,8400	0,6078	3,7832	2,4224	1,5617	0,7948
1,8500	0,6057	3,8263	2,4381	1,5693	0,7902
1,8600	0,6036	3,8695	2,4537	1,5770	0,7857
1,8700	0,6016	3,9131	2,4693	1,5847	0,7811
1,8800	0,5996	3,9568	2,4848	1,5924	0,7765
1,8900	0,5976	4,0008	2,5003	1,6001	0,7720
1,9000	0,5956	4,0450	2,5157	1,6079	0,7674
1,9100	0,5937	4,0895	2,5310	1,6157	0,7627
1,9200	0,5918	4,1341	2,5463	1,6236	0,7581
1,9300	0,5899	4,1791	2,5616	1,6314	0,7535
1,9400	0,5880	4,2242	2,5767	1,6394	0,7488
1,9500	0,5862	4,2696	2,5919	1,6473	0,7442
1,9600	0,5844	4,3152	2,6069	1,6553	0,7395
1,9700	0,5826	4,3611	2,6220	1,6633	0,7349
1,9800	0,5808	4,4071	2,6369	1,6713	0,7302
1,9900	0,5791	4,4535	2,6518	1,6794	0,7255
2,0000	0,5774	4,5000	2,6667	1,6875	0,7209
2,0100	0,5757	4,5468	2,6815	1,6956	0,7162
2,0200	0,5740	4,5938	2,6962	1,7038	0,7115
2,0300	0,5723	4,6411	2,7109	1,7120	0,7069
2,0400	0,5707	4,6885	2,7255	1,7203	0,7022
2,0500	0,5691	4,7363	2,7400	1,7285	0,6975
2,0600	0,5675	4,7842	2,7545	1,7369	0,6928
2,0700	0,5659	4,8324	2,7689	1,7452	0,6882
2,0800	0,5643	4,8808	2,7833	1,7536	0,6835
2,0900	0,5628	4,9295	2,7976	1,7620	0,6789
2,1000	0,5613	4,9783	2,8119	1,7705	0,6742

2,1100	0,5598	5,0275	2,8261	1,7789	0,6696
2,1200	0,5583	5,0768	2,8402	1,7875	0,6649
2,1300	0,5568	5,1264	2,8543	1,7960	0,6603
2,1400	0,5554	5,1762	2,8683	1,8046	0,6557
2,1500	0,5540	5,2263	2,8823	1,8132	0,6511
2,1600	0,5525	5,2765	2,8962	1,8219	0,6464
2,1700	0,5511	5,3271	2,9101	1,8306	0,6419
2,1800	0,5498	5,3778	2,9238	1,8393	0,6373
2,1900	0,5484	5,4288	2,9376	1,8481	0,6327
2,2000	0,5471	5,4800	2,9512	1,8569	0,6281
2,2100	0,5457	5,5315	2,9648	1,8657	0,6236
2,2200	0,5444	5,5831	2,9784	1,8746	0,6191
2,2300	0,5431	5,6351	2,9918	1,8835	0,6145
2,2400	0,5418	5,6872	3,0053	1,8924	0,6100
2,2500	0,5406	5,7396	3,0186	1,9014	0,6055
2,2600	0,5393	5,7922	3,0319	1,9104	0,6011
2,2700	0,5381	5,8451	3,0452	1,9194	0,5966
2,2800	0,5368	5,8981	3,0584	1,9285	0,5921
2,2900	0,5356	5,9515	3,0715	1,9376	0,5877
2,3000	0,5344	6,0050	3,0845	1,9468	0,5833
2,3100	0,5332	6,0588	3,0976	1,9560	0,5789
2,3200	0,5321	6,1128	3,1105	1,9652	0,5745
2,3300	0,5309	6,1671	3,1234	1,9745	0,5702
2,3400	0,5297	6,2215	3,1362	1,9838	0,5658
2,3500	0,5286	6,2763	3,1490	1,9931	0,5615
2,3600	0,5275	6,3312	3,1617	2,0025	0,5572
2,3700	0,5264	6,3864	3,1743	2,0119	0,5529
2,3800	0,5253	6,4418	3,1869	2,0213	0,5486
2,3900	0,5242	6,4975	3,1994	2,0308	0,5444
2,4000	0,5231	6,5533	3,2119	2,0403	0,5401
2,4100	0,5221	6,6095	3,2243	2,0499	0,5359
2,4200	0,5210	6,6658	3,2367	2,0595	0,5317

2,4300	0,5200	6,7224	3,2489	2,0691	0,5276
2,4400	0,5189	6,7792	3,2612	2,0788	0,5234
2,4500	0,5179	6,8363	3,2733	2,0885	0,5193
2,4600	0,5169	6,8935	3,2855	2,0982	0,5152
2,4700	0,5159	6,9511	3,2975	2,1080	0,5111
2,4800	0,5149	7,0088	3,3095	2,1178	0,5071
2,4900	0,5140	7,0668	3,3215	2,1276	0,5030
2,5000	0,5130	7,1250	3,3333	2,1375	0,4990
2,5100	0,5120	7,1835	3,3452	2,1474	0,4950
2,5200	0,5111	7,2421	3,3569	2,1574	0,4911
2,5300	0,5102	7,3011	3,3686	2,1674	0,4871
2,5400	0,5092	7,3602	3,3803	2,1774	0,4832
2,5500	0,5083	7,4196	3,3919	2,1875	0,4793
2,5600	0,5074	7,4792	3,4034	2,1976	0,4754
2,5700	0,5065	7,5391	3,4149	2,2077	0,4715
2,5800	0,5056	7,5991	3,4263	2,2179	0,4677
2,5900	0,5047	7,6595	3,4377	2,2281	0,4639
2,6000	0,5039	7,7200	3,4490	2,2383	0,4601
2,6100	0,5030	7,7808	3,4602	2,2486	0,4564
2,6200	0,5022	7,8418	3,4714	2,2590	0,4526
2,6300	0,5013	7,9031	3,4826	2,2693	0,4489
2,6400	0,5005	7,9645	3,4937	2,2797	0,4452
2,6500	0,4996	8,0263	3,5047	2,2902	0,4416
2,6600	0,4988	8,0882	3,5157	2,3006	0,4379
2,6700	0,4980	8,1504	3,5266	2,3111	0,4343
2,6800	0,4972	8,2128	3,5374	2,3217	0,4307
2,6900	0,4964	8,2755	3,5482	2,3323	0,4271
2,7000	0,4956	8,3383	3,5590	2,3429	0,4236
2,7100	0,4949	8,4015	3,5697	2,3536	0,4201
2,7200	0,4941	8,4648	3,5803	2,3642	0,4166
2,7300	0,4933	8,5284	3,5909	2,3750	0,4131
2,7400	0,4926	8,5922	3,6015	2,3858	0,4097

2,7500	0,4918	8,6563	3,6119	2,3966	0,4062
2,7600	0,4911	8,7205	3,6224	2,4074	0,4028
2,7700	0,4903	8,7851	3,6327	2,4183	0,3994
2,7800	0,4896	8,8498	3,6431	2,4292	0,3961
2,7900	0,4889	8,9148	3,6533	2,4402	0,3928
2,8000	0,4882	8,9800	3,6636	2,4512	0,3895
2,8100	0,4875	9,0455	3,6737	2,4622	0,3862
2,8200	0,4868	9,1111	3,6838	2,4733	0,3829
2,8300	0,4861	9,1771	3,6939	2,4844	0,3797
2,8400	0,4854	9,2432	3,7039	2,4955	0,3765
2,8500	0,4847	9,3096	3,7139	2,5067	0,3733
2,8600	0,4840	9,3762	3,7238	2,5179	0,3701
2,8700	0,4833	9,4431	3,7336	2,5292	0,3670
2,8800	0,4827	9,5101	3,7434	2,5405	0,3639
2,8900	0,4820	9,5775	3,7532	2,5518	0,3608
2,9000	0,4814	9,6450	3,7629	2,5632	0,3577
2,9100	0,4807	9,7128	3,7725	2,5746	0,3547
2,9200	0,4801	9,7808	3,7821	2,5861	0,3517
2,9300	0,4795	9,8491	3,7917	2,5976	0,3487
2,9400	0,4788	9,9175	3,8012	2,6091	0,3457
2,9500	0,4782	9,9863	3,8106	2,6206	0,3428
2,9600	0,4776	10,0552	3,8200	2,6322	0,3398
2,9700	0,4770	10,1244	3,8294	2,6439	0,3369
2,9800	0,4764	10,1938	3,8387	2,6555	0,3340
2,9900	0,4758	10,2635	3,8479	2,6673	0,3312
3,0000	0,4752	10,3333	3,8571	2,6790	0,3283
3,0100	0,4746	10,4035	3,8663	2,6908	0,3255
3,0200	0,4740	10,4738	3,8754	2,7026	0,3227
3,0300	0,4734	10,5444	3,8845	2,7145	0,3200
3,0400	0,4729	10,6152	3,8935	2,7264	0,3172
3,0500	0,4723	10,6863	3,9025	2,7383	0,3145
3,0600	0,4717	10,7575	3,9114	2,7503	0,3118

3,0700	0,4712	10,8291	3,9203	2,7623	0,3091
3,0800	0,4706	10,9008	3,9291	2,7744	0,3065
3,0900	0,4701	10,9728	3,9379	2,7865	0,3038
3,1000	0,4695	11,0450	3,9466	2,7986	0,3012
3,1100	0,4690	11,1175	3,9553	2,8108	0,2986
3,1200	0,4685	11,1901	3,9639	2,8230	0,2960
3,1300	0,4679	11,2631	3,9725	2,8352	0,2935
3,1400	0,4674	11,3362	3,9811	2,8475	0,2910
3,1500	0,4669	11,4096	3,9896	2,8598	0,2885
3,1600	0,4664	11,4832	3,9981	2,8722	0,2860
3,1700	0,4659	11,5571	4,0065	2,8846	0,2835
3,1800	0,4654	11,6311	4,0149	2,8970	0,2811
3,1900	0,4648	11,7055	4,0232	2,9095	0,2786
3,2000	0,4643	11,7800	4,0315	2,9220	0,2762
3,2100	0,4639	11,8548	4,0397	2,9345	0,2738
3,2200	0,4634	11,9298	4,0479	2,9471	0,2715
3,2300	0,4629	12,0051	4,0561	2,9598	0,2691
3,2400	0,4624	12,0805	4,0642	2,9724	0,2668
3,2500	0,4619	12,1563	4,0723	2,9851	0,2645
3,2600	0,4614	12,2322	4,0803	2,9979	0,2622
3,2700	0,4610	12,3084	4,0883	3,0106	0,2600
3,2800	0,4605	12,3848	4,0963	3,0234	0,2577
3,2900	0,4600	12,4615	4,1042	3,0363	0,2555
3,3000	0,4596	12,5383	4,1120	3,0492	0,2533
3,3100	0,4591	12,6155	4,1198	3,0621	0,2511
3,3200	0,4587	12,6928	4,1276	3,0751	0,2489
3,3300	0,4582	12,7704	4,1354	3,0881	0,2468
3,3400	0,4578	12,8482	4,1431	3,1011	0,2446
3,3500	0,4573	12,9263	4,1507	3,1142	0,2425
3,3600	0,4569	13,0045	4,1583	3,1273	0,2404
3,3700	0,4565	13,0831	4,1659	3,1405	0,2383
3,3800	0,4560	13,1618	4,1734	3,1537	0,2363

3,3900	0,4556	13,2408	4,1809	3,1669	0,2342
3,4000	0,4552	13,3200	4,1884	3,1802	0,2322
3,4100	0,4548	13,3995	4,1958	3,1935	0,2302
3,4200	0,4544	13,4791	4,2032	3,2069	0,2282
3,4300	0,4540	13,5591	4,2105	3,2203	0,2263
3,4400	0,4535	13,6392	4,2179	3,2337	0,2243
3,4500	0,4531	13,7196	4,2251	3,2472	0,2224
3,4600	0,4527	13,8002	4,2323	3,2607	0,2205
3,4700	0,4523	13,8811	4,2395	3,2742	0,2186
3,4800	0,4519	13,9621	4,2467	3,2878	0,2167
3,4900	0,4515	14,0435	4,2538	3,3014	0,2148
3,5000	0,4512	14,1250	4,2609	3,3151	0,2129
3,5100	0,4508	14,2068	4,2679	3,3287	0,2111
3,5200	0,4504	14,2888	4,2749	3,3425	0,2093
3,5300	0,4500	14,3711	4,2819	3,3563	0,2075
3,5400	0,4496	14,4535	4,2888	3,3701	0,2057
3,5500	0,4492	14,5363	4,2957	3,3839	0,2039
3,5600	0,4489	14,6192	4,3026	3,3978	0,2022
3,5700	0,4485	14,7024	4,3094	3,4117	0,2004
3,5800	0,4481	14,7858	4,3162	3,4257	0,1987
3,5900	0,4478	14,8695	4,3229	3,4397	0,1970
3,6000	0,4474	14,9533	4,3296	3,4537	0,1953
3,6100	0,4471	15,0375	4,3363	3,4678	0,1936
3,6200	0,4467	15,1218	4,3429	3,4819	0,1920
3,6300	0,4463	15,2064	4,3496	3,4961	0,1903
3,6400	0,4460	15,2912	4,3561	3,5103	0,1887
3,6500	0,4456	15,3763	4,3627	3,5245	0,1871
3,6600	0,4453	15,4615	4,3692	3,5388	0,1855
3,6700	0,4450	15,5471	4,3756	3,5531	0,1839
3,6800	0,4446	15,6328	4,3821	3,5674	0,1823
3,6900	0,4443	15,7188	4,3885	3,5818	0,1807
3,7000	0,4439	15,8050	4,3949	3,5962	0,1792

3,7100	0,4436	15,8915	4,4012	3,6107	0,1777
3,7200	0,4433	15,9781	4,4075	3,6252	0,1761
3,7300	0,4430	16,0651	4,4138	3,6397	0,1746
3,7400	0,4426	16,1522	4,4200	3,6543	0,1731
3,7500	0,4423	16,2396	4,4262	3,6689	0,1717
3,7600	0,4420	16,3272	4,4324	3,6836	0,1702
3,7700	0,4417	16,4151	4,4385	3,6983	0,1687
3,7800	0,4414	16,5031	4,4447	3,7130	0,1673
3,7900	0,4410	16,5915	4,4507	3,7278	0,1659
3,8000	0,4407	16,6800	4,4568	3,7426	0,1645
3,8100	0,4404	16,7688	4,4628	3,7575	0,1631
3,8200	0,4401	16,8578	4,4688	3,7723	0,1617
3,8300	0,4398	16,9471	4,4747	3,7873	0,1603
3,8400	0,4395	17,0365	4,4807	3,8022	0,1589
3,8500	0,4392	17,1263	4,4866	3,8172	0,1576
3,8600	0,4389	17,2162	4,4924	3,8323	0,1563
3,8700	0,4386	17,3064	4,4983	3,8473	0,1549
3,8800	0,4383	17,3968	4,5041	3,8625	0,1536
3,8900	0,4380	17,4875	4,5098	3,8776	0,1523
3,9000	0,4377	17,5783	4,5156	3,8928	0,1510
3,9100	0,4375	17,6695	4,5213	3,9080	0,1497
3,9200	0,4372	17,7608	4,5270	3,9233	0,1485
3,9300	0,4369	17,8524	4,5326	3,9386	0,1472
3,9400	0,4366	17,9442	4,5383	3,9540	0,1460
3,9500	0,4363	18,0363	4,5439	3,9694	0,1448
3,9600	0,4360	18,1285	4,5494	3,9848	0,1435
3,9700	0,4358	18,2211	4,5550	4,0003	0,1423
3,9800	0,4355	18,3138	4,5605	4,0158	0,1411
3,9900	0,4352	18,4068	4,5660	4,0313	0,1399
4,0000	0,4350	18,5000	4,5714	4,0469	0,1388
4,0100	0,4347	18,5935	4,5769	4,0625	0,1376
4,0200	0,4344	18,6871	4,5823	4,0782	0,1364

4,0300	0,4342	18,7811	4,5876	4,0938	0,1353
4,0400	0,4339	18,8752	4,5930	4,1096	0,1342
4,0500	0,4336	18,9696	4,5983	4,1254	0,1330
4,0600	0,4334	19,0642	4,6036	4,1412	0,1319
4,0700	0,4331	19,1591	4,6089	4,1570	0,1308
4,0800	0,4329	19,2541	4,6141	4,1729	0,1297
4,0900	0,4326	19,3495	4,6193	4,1888	0,1286
4,1000	0,4324	19,4450	4,6245	4,2048	0,1276
4,1100	0,4321	19,5408	4,6296	4,2208	0,1265
4,1200	0,4319	19,6368	4,6348	4,2368	0,1254
4,1300	0,4316	19,7331	4,6399	4,2529	0,1244
4,1400	0,4314	19,8295	4,6450	4,2690	0,1234
4,1500	0,4311	19,9263	4,6500	4,2852	0,1223
4,1600	0,4309	20,0232	4,6550	4,3014	0,1213
4,1700	0,4306	20,1204	4,6601	4,3176	0,1203
4,1800	0,4304	20,2178	4,6650	4,3339	0,1193
4,1900	0,4302	20,3155	4,6700	4,3502	0,1183
4,2000	0,4299	20,4133	4,6749	4,3666	0,1173
4,2100	0,4297	20,5115	4,6798	4,3830	0,1164
4,2200	0,4295	20,6098	4,6847	4,3994	0,1154
4,2300	0,4292	20,7084	4,6896	4,4159	0,1144
4,2400	0,4290	20,8072	4,6944	4,4324	0,1135
4,2500	0,4288	20,9063	4,6992	4,4489	0,1126
4,2600	0,4286	21,0055	4,7040	4,4655	0,1116
4,2700	0,4283	21,1051	4,7087	4,4821	0,1107
4,2800	0,4281	21,2048	4,7135	4,4988	0,1098
4,2900	0,4279	21,3048	4,7182	4,5155	0,1089
4,3000	0,4277	21,4050	4,7229	4,5322	0,1080
4,3100	0,4275	21,5055	4,7275	4,5490	0,1071
4,3200	0,4272	21,6061	4,7322	4,5658	0,1062
4,3300	0,4270	21,7071	4,7368	4,5827	0,1054
4,3400	0,4268	21,8082	4,7414	4,5995	0,1045

4,3500	0,4266	21,9096	4,7460	4,6165	0,1036
4,3600	0,4264	22,0112	4,7505	4,6334	0,1028
4,3700	0,4262	22,1131	4,7550	4,6505	0,1020
4,3800	0,4260	22,2151	4,7595	4,6675	0,1011
4,3900	0,4258	22,3175	4,7640	4,6846	0,1003
4,4000	0,4255	22,4200	4,7685	4,7017	0,0995
4,4100	0,4253	22,5228	4,7729	4,7189	0,0987
4,4200	0,4251	22,6258	4,7773	4,7361	0,0979
4,4300	0,4249	22,7291	4,7817	4,7533	0,0971
4,4400	0,4247	22,8325	4,7861	4,7706	0,0963
4,4500	0,4245	22,9363	4,7904	4,7879	0,0955
4,4600	0,4243	23,0402	4,7948	4,8053	0,0947
4,4700	0,4241	23,1444	4,7991	4,8227	0,0940
4,4800	0,4239	23,2488	4,8034	4,8401	0,0932
4,4900	0,4237	23,3535	4,8076	4,8576	0,0924
4,5000	0,4236	23,4583	4,8119	4,8751	0,0917
4,5100	0,4234	23,5635	4,8161	4,8926	0,0910
4,5200	0,4232	23,6688	4,8203	4,9102	0,0902
4,5300	0,4230	23,7744	4,8245	4,9279	0,0895
4,5400	0,4228	23,8802	4,8287	4,9455	0,0888
4,5500	0,4226	23,9863	4,8328	4,9632	0,0881
4,5600	0,4224	24,0925	4,8369	4,9810	0,0874
4,5700	0,4222	24,1991	4,8410	4,9987	0,0867
4,5800	0,4220	24,3058	4,8451	5,0166	0,0860
4,5900	0,4219	24,4128	4,8492	5,0344	0,0853
4,6000	0,4217	24,5200	4,8532	5,0523	0,0846
4,6100	0,4215	24,6275	4,8572	5,0703	0,0839
4,6200	0,4213	24,7351	4,8612	5,0882	0,0832
4,6300	0,4211	24,8431	4,8652	5,1063	0,0826
4,6400	0,4210	24,9512	4,8692	5,1243	0,0819
4,6500	0,4208	25,0596	4,8731	5,1424	0,0813
4,6600	0,4206	25,1682	4,8771	5,1605	0,0806

4,6700	0,4204	25,2771	4,8810	5,1787	0,0800
4,6800	0,4203	25,3861	4,8849	5,1969	0,0793
4,6900	0,4201	25,4955	4,8887	5,2151	0,0787
4,7000	0,4199	25,6050	4,8926	5,2334	0,0781
4,7100	0,4197	25,7148	4,8964	5,2518	0,0775
4,7200	0,4196	25,8248	4,9002	5,2701	0,0769
4,7300	0,4194	25,9351	4,9040	5,2885	0,0762
4,7400	0,4192	26,0455	4,9078	5,3070	0,0756
4,7500	0,4191	26,1563	4,9116	5,3254	0,0750
4,7600	0,4189	26,2672	4,9153	5,3440	0,0745
4,7700	0,4187	26,3784	4,9190	5,3625	0,0739
4,7800	0,4186	26,4898	4,9227	5,3811	0,0733
4,7900	0,4184	26,6015	4,9264	5,3997	0,0727
4,8000	0,4183	26,7133	4,9301	5,4184	0,0721
4,8100	0,4181	26,8255	4,9338	5,4371	0,0716
4,8200	0,4179	26,9378	4,9374	5,4559	0,0710
4,8300	0,4178	27,0504	4,9410	5,4747	0,0705
4,8400	0,4176	27,1632	4,9446	5,4935	0,0699
4,8500	0,4175	27,2763	4,9482	5,5124	0,0694
4,8600	0,4173	27,3895	4,9518	5,5313	0,0688
4,8700	0,4172	27,5031	4,9553	5,5502	0,0683
4,8800	0,4170	27,6168	4,9589	5,5692	0,0677
4,8900	0,4169	27,7308	4,9624	5,5882	0,0672
4,9000	0,4167	27,8450	4,9659	5,6073	0,0667
4,9100	0,4165	27,9595	4,9694	5,6264	0,0662
4,9200	0,4164	28,0741	4,9728	5,6455	0,0657
4,9300	0,4163	28,1891	4,9763	5,6647	0,0652
4,9400	0,4161	28,3042	4,9797	5,6839	0,0647
4,9500	0,4160	28,4196	4,9831	5,7032	0,0642
4,9600	0,4158	28,5352	4,9865	5,7224	0,0637
4,9700	0,4157	28,6511	4,9899	5,7418	0,0632
4,9800	0,4155	28,7671	4,9933	5,7611	0,0627

4,9900	0,4154	28,8835	4,9967	5,7806	0,0622
5,0000	0,4152	29,0000	5,0000	5,8000	0,0617
5,0100	0,4151	29,1168	5,0033	5,8195	0,0612
5,0200	0,4149	29,2338	5,0066	5,8390	0,0608
5,0300	0,4148	29,3511	5,0099	5,8586	0,0603
5,0400	0,4147	29,4685	5,0132	5,8782	0,0598
5,0500	0,4145	29,5863	5,0165	5,8978	0,0594
5,0600	0,4144	29,7042	5,0197	5,9175	0,0589
5,0700	0,4142	29,8224	5,0230	5,9372	0,0585
5,0800	0,4141	29,9408	5,0262	5,9570	0,0580
5,0900	0,4140	30,0595	5,0294	5,9768	0,0576
5,1000	0,4138	30,1783	5,0326	5,9966	0,0572
5,1100	0,4137	30,2975	5,0357	6,0165	0,0567
5,1200	0,4136	30,4168	5,0389	6,0364	0,0563
5,1300	0,4134	30,5364	5,0421	6,0563	0,0559
5,1400	0,4133	30,6562	5,0452	6,0763	0,0554
5,1500	0,4132	30,7763	5,0483	6,0964	0,0550
5,1600	0,4130	30,8965	5,0514	6,1164	0,0546
5,1700	0,4129	31,0171	5,0545	6,1365	0,0542
5,1800	0,4128	31,1378	5,0576	6,1567	0,0538
5,1900	0,4126	31,2588	5,0606	6,1769	0,0534
5,2000	0,4125	31,3800	5,0637	6,1971	0,0530
5,2100	0,4124	31,5015	5,0667	6,2173	0,0526
5,2200	0,4123	31,6231	5,0697	6,2376	0,0522
5,2300	0,4121	31,7451	5,0727	6,2580	0,0518
5,2400	0,4120	31,8672	5,0757	6,2784	0,0514
5,2500	0,4119	31,9896	5,0787	6,2988	0,0510
5,2600	0,4118	32,1122	5,0817	6,3192	0,0506
5,2700	0,4116	32,2351	5,0846	6,3397	0,0502
5,2800	0,4115	32,3581	5,0875	6,3603	0,0499
5,2900	0,4114	32,4815	5,0905	6,3808	0,0495
5,3000	0,4113	32,6050	5,0934	6,4014	0,0491

5,3100	0,4112	32,7288	5,0963	6,4221	0,0488
5,3200	0,4110	32,8528	5,0992	6,4428	0,0484
5,3300	0,4109	32,9771	5,1020	6,4635	0,0480
5,3400	0,4108	33,1015	5,1049	6,4843	0,0477
5,3500	0,4107	33,2263	5,1077	6,5051	0,0473
5,3600	0,4106	33,3512	5,1106	6,5259	0,0470
5,3700	0,4104	33,4764	5,1134	6,5468	0,0466
5,3800	0,4103	33,6018	5,1162	6,5677	0,0463
5,3900	0,4102	33,7275	5,1190	6,5887	0,0459
5,4000	0,4101	33,8533	5,1218	6,6097	0,0456
5,4100	0,4100	33,9795	5,1246	6,6307	0,0453
5,4200	0,4099	34,1058	5,1273	6,6518	0,0449
5,4300	0,4098	34,2324	5,1301	6,6729	0,0446
5,4400	0,4096	34,3592	5,1328	6,6941	0,0443
5,4500	0,4095	34,4863	5,1355	6,7153	0,0439
5,4600	0,4094	34,6135	5,1382	6,7365	0,0436
5,4700	0,4093	34,7411	5,1409	6,7578	0,0433
5,4800	0,4092	34,8688	5,1436	6,7791	0,0430
5,4900	0,4091	34,9968	5,1463	6,8004	0,0427
5,5000	0,4090	35,1250	5,1489	6,8218	0,0424
5,5100	0,4089	35,2535	5,1516	6,8432	0,0421
5,5200	0,4088	35,3821	5,1542	6,8647	0,0417
5,5300	0,4086	35,5111	5,1569	6,8862	0,0414
5,5400	0,4085	35,6402	5,1595	6,9077	0,0411
5,5500	0,4084	35,7696	5,1621	6,9293	0,0408
5,5600	0,4083	35,8992	5,1647	6,9509	0,0405
5,5700	0,4082	36,0291	5,1672	6,9726	0,0403
5,5800	0,4081	36,1591	5,1698	6,9943	0,0400
5,5900	0,4080	36,2895	5,1724	7,0160	0,0397
5,6000	0,4079	36,4200	5,1749	7,0378	0,0394
5,6100	0,4078	36,5508	5,1775	7,0596	0,0391
5,6200	0,4077	36,6818	5,1800	7,0815	0,0388

5,6300	0,4076	36,8131	5,1825	7,1033	0,0385
5,6400	0,4075	36,9445	5,1850	7,1253	0,0383
5,6500	0,4074	37,0763	5,1875	7,1472	0,0380
5,6600	0,4073	37,2082	5,1900	7,1693	0,0377
5,6700	0,4072	37,3404	5,1924	7,1913	0,0374
5,6800	0,4071	37,4728	5,1949	7,2134	0,0372
5,6900	0,4070	37,6055	5,1973	7,2355	0,0369
5,7000	0,4069	37,7383	5,1998	7,2577	0,0366
5,7100	0,4068	37,8715	5,2022	7,2799	0,0364
5,7200	0,4067	38,0048	5,2046	7,3021	0,0361
5,7300	0,4066	38,1384	5,2070	7,3244	0,0359
5,7400	0,4065	38,2722	5,2094	7,3467	0,0356
5,7500	0,4064	38,4063	5,2118	7,3691	0,0354
5,7600	0,4063	38,5405	5,2142	7,3915	0,0351
5,7700	0,4062	38,6751	5,2166	7,4139	0,0349
5,7800	0,4061	38,8098	5,2189	7,4364	0,0346
5,7900	0,4060	38,9448	5,2213	7,4589	0,0344
5,8000	0,4059	39,0800	5,2236	7,4814	0,0341
5,8100	0,4059	39,2155	5,2259	7,5040	0,0339
5,8200	0,4058	39,3511	5,2282	7,5266	0,0336
5,8300	0,4057	39,4871	5,2306	7,5493	0,0334
5,8400	0,4056	39,6232	5,2328	7,5720	0,0332
5,8500	0,4055	39,7596	5,2351	7,5948	0,0329
5,8600	0,4054	39,8962	5,2374	7,6175	0,0327
5,8700	0,4053	40,0331	5,2397	7,6404	0,0325
5,8800	0,4052	40,1701	5,2419	7,6632	0,0322
5,8900	0,4051	40,3075	5,2442	7,6861	0,0320
5,9000	0,4050	40,4450	5,2464	7,7091	0,0318
5,9100	0,4049	40,5828	5,2486	7,7320	0,0316
5,9200	0,4049	40,7208	5,2509	7,7551	0,0314
5,9300	0,4048	40,8591	5,2531	7,7781	0,0311
5,9400	0,4047	40,9975	5,2553	7,8012	0,0309

5,9500	0,4046	41,1363	5,2575	7,8243	0,0307
5,9600	0,4045	41,2752	5,2597	7,8475	0,0305
5,9700	0,4044	41,4144	5,2618	7,8707	0,0303
5,9800	0,4043	41,5538	5,2640	7,8940	0,0301
5,9900	0,4042	41,6935	5,2661	7,9173	0,0299
6,0000	0,4042	41,8333	5,2683	7,9406	0,0297

Bibliography

[ALM 13] ALMOHAMMADI K.M., INGHAM D.B., MA L. *et al.*, "Computational fluid dynamics (CFD) mesh independency techniques for a straight blade vertical axis wind turbine", *Energy*, vol. 58, no. 1, pp. 483–493, 2013.

[ANS 09] ANSYS Inc., *ANSYS FLUENT 12.0 Manual*, ANSYS Inc., 2009.

[ANS 14] ANSYS USER GUIDE, Sliding Mesh Theory, Chapter 3: Flows Using Sliding and Dynamic Meshes, 3.3, 2014.

[BEN 08] BENDAOU O., Réduction de modèle des problèmes Hydro-Elastiques à Paramètres Incertains, Thesis, INSA Rouen, 2008.

[BEN 14] BEN ABDESSALEM A., EL-HAMI A., "Global sensitivity analysis and multi-objective optimisation of loading path in tube hydroforming process based on metamodelling techniques", *International Journal of Advanced Manufacturing Technology*, vol. 71, nos. 5–8, pp. 753–773, 2014.

[B0R 07] BORZA D., BENDAOU O., MAKHLOUFI A. *et al.*, "Numerical and experimental Analysis of Hydro-Elastic Problems", *Eleventh International Conference on Civil, Structural and Environmental Engineering Computing*, St. Julians, Malta, pp. 18–21 September, 2007.

[CAI 11] CAI-WAN CHANG-JIAN, "Nonlinear dynamic analysis for bevel-gear system under nonlinear suspension-bifurcation and chaos", *Applied Mathematical Modeling*, vol. 35, pp. 3225–3237, 2011.

[CAR 12] CARRIGAN T.J., DENNIS B.H., HAN Z.X. *et al.*, "Aerodynamic shape optimization of a vertical-axis wind turbine using differential evolution", *ISRN Renewable Energy*, 2012.

[CON 11] CONSUL C.A., Hydrodynamic analysis of a tidal cross-flow turbine, Thesis, University of Oxford, 2011.

[DUR 11] DURRANI N., HAMEED H., RAHMAN H. *et al.*, "A detailed aerodynamic design and analysis of a 2D vertical axis wind turbine using sliding mesh in CFD", *49th AIAA Aerospace Sciences Meeting*, Orlando, Florida, 4–7 January 2011.

[EBO 13] EBOIBI O., The influence of blade chord on the aerodynamics and performance of vertical axis wind turbines, PhD Thesis, University of Sheffield, 2013.

[EL 96] EL HAMI A., RADI B., "Some decomposition methods in the analysis of repetitive structures", *International Journal Computers & Structures*, vol. 58, no. 5, pp. 973–980, 1996.

[EL 11] EL HAMI A., RADI B., "Comparison study of different reliability-based design optimization approaches", *Advanced Materials Research*, vol. 274, pp. 119–130, 2011.

[EL 13] EL HAMI A., RADI B., *Incertitude et optimisation et fiabilité des structures*, Hermes Science-Lavoisier, Paris, 2013.

[ELH 93] ELHAMI A., LALLEMENT G., MINOTTI P. *et al.*, "Methods that combine finite group theory with component mode synthesis in the analysis of repetitive structures", *International Journal Computers & Structures*, vol. 48, no. 6, pp. 975–982, 1993.

[ELS 10] EL-SAMANOUDY E., GHORAB A., YOUSSEF S.Z., "Effect of some design parameters on the performance of a giromill vertical axis wind turbine", *Ain Shams Engineering Journal*, vol. 1, pp. 85–95, 2010.

[FAZ 14] FAZIL S.G., Performance prediction of Darrieus type vertical axis wind turbines using numerical simulations, PhD Thesis, 2014.

[FLU 06] FLUENT INC., FLUENT 6.3 User's Guide, September 2006.

[GOS 13] GOSSELIN R., DUMAS G., BOUDREAU M., "Parametric study of H-Darrieus vertical-axis turbines using URANS simulations", *CFDSC*, 2013.

[GUE 15a] GUERINE A., EL HAMI A., WALHA L. *et al.*, "A perturbation approach for the dynamic analysis of one stage gear system with uncertain parameters", *Mechanism and Machine Theory*, vol. 92, pp. 113–126, 2015.

[GUE 15b] GUERINE A., EL HAMI A., FAKHFAKH T. *et al.*, "A polynomial chaos method to the analysis of the dynamic behavior of spur gear system", *Structural Engineering and Mechanics An International Journal*, vol. 53, pp. 819–831, 2015.

[GUE 16] GUERRINE A., Contribution à l'étude de comportement dynamique d'un train épicycloïdal en présence d'incertitudes, Thesis, INSA Rouen, 2016.

[HAS 14] HASSAN G.E., HASSAN A., YOUSSEF M.E., "Numerical investigation of medium range Re number aerodynamics characteristics for NACA0018 airfoil", *CFD Letters*, vol. 6, no. 4, pp. 175–187, 2014.

[ION 14] ION M., HORIA D., "Numerical simulation of VAWT flow using fluent", *UPB Scientific Bulletin, Series D: Mechanical Engineering*, vol. 76, no. 1, pp. 109–122, 2014.

[IRI 07] IRISO J.G., Analysis and design of a vertical axis wind turbine, Thesis, ETSIIT, 2007.

[ISL 08] ISLAM M., FARTAJ A., CARRIVEAU R., "Analysis of the design parameters related to a fixed-pitch straight-bladed vertical axis wind turbine", *Wind Engineering*, vol. 32, no. 5, pp. 491–507, 2008.

[JAM 01] JAMATI F., Etude numérique d'une éolienne hybride asynchrone, Thesis, Montreal University, 2001.

[JON 13] JONATHAN M.E., LOUIS A.D., ROBERT J.H., "PIV measurements and CFD simulation of the performance and flow physics and of a small-scale vertical axis wind turbine", *Wind Energy*, vol. 18, pp. 201–217, 2013.

[LED 17] LEDOUX M., EL HAMI A., *Fluid Mechanics*, ISTE, London and John Wiley & Sons, New York, 2017.

[MAK 08] MAKHLOUFI A., Contribution aux développements et à la validation des méthodes de l'optimisation fiabiliste de conception en dynamique des structures mécaniques, Thesis, INSA Rouen, 2008.

[MEN 94] MENTER F.R., "Two-equation eddy-viscosity turbulence models for engineering applications", *AIAA Journal*, vol. 32, no. 8, pp. 1598–1605, 1994.

[MOH 03] MOHSINE A., KHARMANDA G., EL HAMI A., "Integration of equivalent model within reliability-based design optimization for reducing freedom degrees in dynamic structures", *Proceedings of 11th IFIP Conferences on Reliability and Optimization Systems*, Canada, pp. 157–164, 2003.

[MOH 05] MOHSINE A., KHARMANDA G., EL HAMI A., "Improved hybrid method for reliability-based design optimization", *12th WG 7.5 Working Conference on Reliability and Optimization of Structural Systems*, Aalborg, Denmark, 22–25 May 2005.

[MOH 10] MOHSINE A., EL HAMI A., "A robust study of reliability-based optimisation methods under eigen-frequency", *International Journal of Computer Methods in Applied Mechanics and Engineering*, vol. 199, no. 17–20, pp. 1006–1018, 2010.

[MOH 12] MOHAMED M.H., "Performance investigation of H-rotor Darrieus turbine with new airfoil shapes", *Energy*, vol. 47, pp. 522–530, 2012.

[NOB 11] NOBILE R., VAHDATI M., BARLOW, J.F. *et al.*, "Dynamic stall for a vertical axis wind turbine in a two-dimensional study", *World Renewable Energy Congress 2011 – Sweden*, Linköping, Sweden, 2011.

[NOB 14] NOBILE R., VAHDATI M., BARLOW J.F. *et al.*, "Unsteady flow simulation of a vertical axis augmented wind turbine: a two-dimensional study", *Journal of Wind Engineering and Industrial Aerodynamics*, vol. 125, pp. 168–179, 2014.

[PAR 02] PARASCHIVOIU I., *Wind Turbine Design: with Emphasis on Darrieus Concept*, Polytechnique International Presses, Montreal, Canada, 2002.

[PAY 08] PAYETTE F.-A., Simulation de l'écoulement turbulent dans les aspirateurs de turbines hydrauliques: Impact des paramètres de modélisation, Thesis, Quebec, 2008.

[PRY 89] PRYPUTNIEWICZ R.J., STETSON K.A., "Measurement of vibration patterns using electro-optic holography", *Proc. SPIE*, no. 1162, pp. 456–467, 1989.

[RAC 11] RACITI CASTELLI M., ENGLARO A., BENIN E., "The Darrieus wind turbine: proposal for a new performance prediction model based on CFD", *Energy*, vol. 36, pp. 4919–4934, 2011.

[RAD 10] RADI B., EL HAMI A., "The study of the dynamic contact in ultrasonic motor", *Applied Mathematical Modelling*, vol. 34, no. 12, pp. 3767–3777, 2010.

[ROJ 10] ROJAS E., BENDAOU O., EL HAMI A. *et al.*, "Stochastic and reliability analysis of fluid–structure interaction problems using finite element models", *International Journal of Multidiscipline Modeling in Materials and Structures*, vol. 6, no. 1, pp. 6–22, 2010.

[ROS 14] ROSARIO L., STEFANO M., MICHELE M., "2D CFD modeling of H-Darrieus wind turbines using a transition turbulence model", *Energy Procedia*, vol. 45, pp. 131–140, 2014.

[SAL 15] SALMAN SIDDIQUI M., DURRANI N., AKHTAR I., "Quantification of the effects of geometric approximations on the performance of a vertical axis wind turbine", *Renewable Energy*, vol. 74, pp. 661–670, 2015.

[SAR 04] SARSRI D., Contribution à l'analyse des structures comportant des paramètres incertains par les méthodes de synthèse modale, Thesis, INSA Rouen, France, 2004.

[SAR 11] SARSRI D., AZRAR L., JEBBOURI A. *et al.*, "Component mode synthesis and polynomial chaos expansions for stochastic frequency functions of large linear FE models", *Computers & Structures*, vol. 89, pp. 346–356, 2011.

[SCH 07] SCHMITT F.G., "About Boussinesq's turbulent viscosity hypothesis: historical remarks and a direct evaluation of its validity", *Comptes Rendus M'ecanique*, vol. 335, pp. 617–627, 2007.

[SID 15] SIDDIQUI M.S., NAVEED D., IMRAN A., "Quantification of effects of geometric approximations on the performance of a vertical axis wind turbine", *Renewable Energy*, vol. 74, pp. 661–670, 2015.

[STE 98] STETSON K.A., WAHID J., "Real-time Phase Imaging for Nondestructive Testing", *Exp. Tech.*, vol. 22, pp. 15–18, 1998.

[TAN 00] TANGLER J.L., The evolution of rotor and blade design, Technical Report, NREL/CP-500-28410, 2000.

[VAI 10] VAISHNAV E., An investigation on the aerodynamic performance of a vertical axis wind turbine, Thesis, Bhilai Institute of Technology, December 2010.

Index

A, B, C

D, E, F

H, I, M

N, O, R

S, T, V

Other titles from

iSTE

in

Mechanical Engineering and Solid Mechanics

2016

BOREL Michel, VÉNIZÉLOS Georges
Movement Equations 1: Location, Kinematics and Kinetics
(Non-deformable Solid Mechanics Set – Volume 1)

Movement Equations 2: Mathematical and Methodological Supplements
(Non-deformable Solid Mechanics Set – Volume 2)

BOYARD Nicolas
Heat Transfer in Polymer Composite Materials

CARDON Alain, ITMI Mhamed
New Autonomous Systems
(Reliability of Multiphysical Systems Set – Volume 1)

DAHOO Pierre Richard, POUGNET Philippe, EL HAMI Abdelkhalak
Nanometer-scale Defect Detection Using Polarized Light
(Reliability of Multiphysical Systems Set – Volume 2)

DE SAXCÉ Géry, VALLÉE Claude
Galilean Mechanics and Thermodynamics of Continua

DORMIEUX Luc, KONDO Djimédo
Micromechanics of Fracture and Damage (Micromechanics Set – Volume 1)

EL HAMI Abdelkhalak, RADI Bouchaib
Stochastic Dynamics of Structures (Mathematical and Mechanical Engineering Set – Volume 2)

GOURIVEAU Rafael, MEDJAHER Kamal, ZERHOUNI Noureddine
From Prognostics and Health Systems Management to Predictive Maintenance 1: Monitoring and Prognostics (Reliability of Multiphysical Systems Set – Volume 4)

KHARMANDA Ghias, EL HAMI Abdelkhalak
Reliability in Biomechanics (Reliability of Multiphysical Systems Set – Volume 3)

Biomechanics: Optimization, Uncertainties and Reliability (Reliability of Multiphysical Systems Set – Volume 5)

MOLIMARD Jérôme
Experimental Mechanics of Solids and Structures

RADI Bouchaib, EL HAMI Abdelkhalak
Material Forming Processes: Simulation, Drawing, Hydroforming and Additive Manufacturing (Mathematical and Mechanical Engineering Set – Volume 1)

2015

KARLIČIĆ Danilo, MURMU Tony, ADHIKARI Sondipon, MCCARTHY Michael
Non-local Structural Mechanics

SAB Karam, LEBÉE Arthur
Homogenization of Heterogeneous Thin and Thick Plates

2014

ATANACKOVIC M. Teodor, PILIPOVIC Stevan, STANKOVIC Bogoljub, ZORICA Dusan
Fractional Calculus with Applications in Mechanics: Vibrations and Diffusion Processes

ATANACKOVIC M. Teodor, PILIPOVIC Stevan, STANKOVIC Bogoljub, ZORICA Dusan
Fractional Calculus with Applications in Mechanics: Wave Propagation, Impact and Variational Principles

CIBLAC Thierry, MOREL Jean-Claude
Sustainable Masonry: Stability and Behavior of Structures

ILANKO Sinniah, MONTERRUBIO Luis E., MOCHIDA Yusuke
The Rayleigh–Ritz Method for Structural Analysis

LALANNE Christian
Mechanical Vibration and Shock Analysis – 5-volume series – 3rd edition
Sinusoidal Vibration – volume 1
Mechanical Shock – volume 2
Random Vibration – volume 3
Fatigue Damage – volume 4
Specification Development – volume 5

LEMAIRE Maurice
Uncertainty and Mechanics

2013

ADHIKARI Sondipon
Structural Dynamic Analysis with Generalized Damping Models: Analysis

ADHIKARI Sondipon
Structural Dynamic Analysis with Generalized Damping Models: Identification

BAILLY Patrice
Materials and Structures under Shock and Impact

BASTIEN Jérôme, BERNARDIN Frédéric, LAMARQUE Claude-Henri
Non-smooth Deterministic or Stochastic Discrete Dynamical Systems: Applications to Models with Friction or Impact

EL HAMI Abdelkhalak, BOUCHAIB Radi
Uncertainty and Optimization in Structural Mechanics

KIRILLOV Oleg N., PELINOVSKY Dmitry E.
Nonlinear Physical Systems: Spectral Analysis, Stability and Bifurcations

LUONGO Angelo, ZULLI Daniele
Mathematical Models of Beams and Cables

SALENÇON Jean
Yield Design

2012

DAVIM J. Paulo
Mechanical Engineering Education

DUPEUX Michel, BRACCINI Muriel
Mechanics of Solid Interfaces

ELISHAKOFF Isaac *et al.*
Carbon Nanotubes and Nanosensors: Vibration, Buckling and Ballistic Impact

GRÉDIAC Michel, HILD François
Full-Field Measurements and Identification in Solid Mechanics

GROUS Ammar
Fracture Mechanics – 3-volume series
Analysis of Reliability and Quality Control – volume 1
Applied Reliability – volume 2
Applied Quality Control – volume 3

RECHO Naman
Fracture Mechanics and Crack Growth

2011

KRYSINSKI Tomasz, MALBURET François
Mechanical Instability

SOUSTELLE Michel
An Introduction to Chemical Kinetics

2010

BREITKOPF Piotr, FILOMENO COELHO Rajan
Multidisciplinary Design Optimization in Computational Mechanics

DAVIM J. Paulo
Biotribolgy

PAULTRE Patrick
Dynamics of Structures

SOUSTELLE Michel
Handbook of Heterogenous Kinetics

2009

BERLIOZ Alain, TROMPETTE Philippe
Solid Mechanics using the Finite Element Method

LEMAIRE Maurice
Structural Reliability

2007

GIRARD Alain, ROY Nicolas
Structural Dynamics in Industry

GUINEBRETIÈRE René
X-ray Diffraction by Polycrystalline Materials

KRYSINSKI Tomasz, MALBURET François
Mechanical Vibrations

KUNDU Tribikram
Advanced Ultrasonic Methods for Material and Structure Inspection

SIH George C. *et al.*
Particle and Continuum Aspects of Mesomechanics

www.ingramcontent.com/pod-product-compliance
Lightning Source LLC
LaVergne TN
LVHW020850250626
841902LV00003B/28